AF410863

Selected Papers from the ICEUBI2019 – International Congress on Engineering – Engineering for Evolution

Selected Papers from the ICEUBI2019 – International Congress on Engineering – Engineering for Evolution

Editors

Maria do Rosário Alves Calado
Jorge Miguel dos Reis Silva

MDPI • Basel • Beijing • Wuhan • Barcelona • Belgrade • Manchester • Tokyo • Cluj • Tianjin

Editors
Maria do Rosário Alves Calado   Jorge Miguel dos Reis Silva
University of Beira Interior   University of Beira Interior
Portugal   Portugal

Editorial Office
MDPI
St. Alban-Anlage 66
4052 Basel, Switzerland

This is a reprint of articles from the Special Issue published online in the open access journal *Energies* (ISSN 1996-1073) (available at: https://www.mdpi.com/journal/energies/special_issues/ICEUBI_2019_Energies).

For citation purposes, cite each article independently as indicated on the article page online and as indicated below:

LastName, A.A.; LastName, B.B.; LastName, C.C. Article Title. *Journal Name* **Year**, *Volume Number*, Page Range.

ISBN 978-3-0365-0668-5 (Hbk)
ISBN 978-3-0365-0669-2 (PDF)

Cover image courtesy of Faculty of Engineering, University of Beira Interior.

Contents

About the Editors

Maria do Rosário Alves Calado is currently Assistant Professor with the Department of Electromechanical Engineering, University of Beira Interior, and a researcher at the Instituto de Telecomunicações (IT), Covilhã, Portugal. She received her electrical engineering degree from the Instituto Superior Técnico (IST), Lisbon, Portugal, in 1991, and her MSc equivalent degree, her PhD degree and habilitation from University of Beira Interior, Covilhã, Portugal, in 1996, 2002 and 2019, respectively. Her research interests include power systems optimization and control, energy conversion systems, renewable energy and energy harvesting and electric road vehicles. She has about a hundred scientific publications.

Jorge Miguel dos Reis Silva graduated in Electronics and Telecommunications Marine Systems Engineering. He has an MSc in Operations Research and Systems Engineering, and a PhD in Transportation. He has been professionally active since 1979: Merchant Navy Officer (1979/1997), Aeronautical Telecommunications Technician (1988/1995), and Professor at University of Beira Interior (UBI), (since 1995). He is Assistant Professor at the Aerospace Sciences Department (DCA) of UBI, and Full Researcher at the Institute for Research and Innovation in Civil Engineering for Sustainability (CERis). He was President of Scientific Council of DCA of UBI between 2006/2009. His scientific activity interests and areas are Economy and Management of Air Transport, Aircraft Operations, Flight Safety/Security, and Operations Research.

Preface to "Selected Papers from the ICEUBI2019 – International Congress on Engineering – Engineering for Evolution"

This *Energies* Special Issue Book "Elected Papers from the ICEUBI2019 – International Congress on Engineering – Engineering for Evolution" comprises six papers that are the latest advances in basic and application research in the field of engineering, specifically covering areas of aeronautics, electrotechnics and mechanics. The accepted papers illustrate the highly innovative and informative venue for essential and advanced scientific and engineering research in those fields. Magalhães et al. propose a Reynolds averaged Navier–Stokes (RANS) computational method following an incompressible but a variable density approach by to the study of supercritical nitrogen mixing layers, in which the performances of several turbulence models are compared in conjunction with a high accuracy multi-parameter equation of state. In addition, a suitable methodology to describe transport properties accounting for dense fluid corrections is applied. The results are discussed and validated against experimental data. Considerations about the applicability of the tested turbulence models in supercritical simulations are given based on the results and each model's structural nature. Camacho et al. present the computational study of the influence of the Reynolds number, frequency, and amplitude of the oscillatory movement of a NACA0012 airfoil in the aerodynamic performance. The thrust and power coefficients are obtained, which together are used to calculate the propulsive efficiency. The simulations were performed using ANSYS Fluent with a RANS approach for Reynolds numbers between 8500 and 34,000, reduced frequencies between 1 and 5, and Strouhal numbers from 0.1 to 0.4. The aerodynamic parameters and their interaction are explored and discussed. Alves et al. give a brief overview of aviation performance trends since the past century. Comparison examples between aircrafts designed in different paradigms are presented. Cruising speed trends and other performance parameters are discussed, looking at different periods and circumstances. The use of aircraft propellers as a reborn propulsive device and its role in aviation progress, even for commercial aviation, is discussed. New playgrounds for propeller innovation like the electric VTOL aircraft aimed for urban mobility are presented. Faria et al. propose a new method for the simultaneous determination of the optimal control parameters of proportional resonant controllers and the optimal design of the output filter of a grid-tied three-phase inverter, based on the grey wolf optimization (GWO) algorithm. The proposed optimization methodology is validated, by using two output filter topologies with series and complex passive damping methods. Bento et al. propose a novel Lagrangian multiplier update adaptative algorithm to automatically adjust the step-size used to update Lagrange multipliers. This new adaptative step-size update is a crucial stage when employing Lagrangian relaxation to solve the hydro-thermal coordination problem in a conventional electrical power generation system. This coordination problem is intrinsically complex, with a large-scale and constrained in nature; thus, the feasibility of a direct approach is reduced. Hence, Lagrangian relaxation, a decomposition method, is a consolidated choice to "simplify" the problem by deriving and solving the associated dual problem. A results comparison is made against two traditionally employed step-size update heuristics, using two real hydrothermal scenarios derived from the Portuguese power system. Santos et al., knowing the unique and growing importance of water, propose a new index, total water impact (TWI), which allows a holistic comparison of the impact of water use on water, air and evaporative condensation climate systems. Energy demand growing for air conditioning comes from a combination of rising temperatures, rising population and economic growth. This increase in energy will directly impact water consumption, either to directly cool the condenser of equipment or to serve indirectly as a basis for energy sources. The proposed index provides a new insight about energy consumption and ultimately, about sustainability. To conclude, the Special Issue editors would like to thank those who have contributed to this book for their high-quality submissions. We would also like to thank those who performed reviews of the manuscripts and their valuable feedback.

Maria do Rosário Alves Calado, Jorge Miguel dos Reis Silva
Editors

Article

Turbulence Modeling Insights into Supercritical Nitrogen Mixing Layers

Leandro Magalhães *, Francisco Carvalho, André Silva and Jorge Barata

LAETA-Aeronautics and Astronautics Research Center, University of Beira Interior, 6201-001 Covilhã, Portugal; francisco.carvalho@ubi.pt (F.C.); andre@ubi.pt (A.S.); jbarata@ubi.pt (J.B.)
* Correspondence:leandro.magalhaes@ubi.pt

Received:27 February 2020; Accepted: 28 March 2020; Published: 1 April 2020

Abstract: In Liquid Rocket Engines, higher combustion efficiencies come at the cost of the propellants exceeding their critical point conditions and entering the supercritical domain. The term fluid is used because, under these conditions, there is no longer a clear distinction between a liquid and a gas phase. The non-conventional behavior of thermophysical properties makes the modeling of supercritical fluid flows a most challenging task. In the present work, a Reynolds Averaged Navier Stokes (RANS) computational method following an incompressible but variable density approach is devised on which the performance of several turbulence models is compared in conjunction with a high accuracy multi-parameter equation of state. In addition, a suitable methodology to describe transport properties accounting for dense fluid corrections is applied. The results are validated against experimental data, making it clear that there is no trend between turbulence model complexity and the quality of the produced results. For several instances, one- and two-equation turbulence models produce similar results. Finally, considerations about the applicability of the tested turbulence models in supercritical simulations are given based on the results and the structural nature of each model.

Keywords: turbulence modeling; supercritical injection; Liquid Rocket Engines

---

1. Introduction

A small step towards the validation of numerical solvers able to accurately replicate the behavior of supercritical fluid flows is to understand how current RANS (Reynolds Averaged Navier Stokes) turbulence models, calibrated and tested for subcritical conditions, behave in the supercritical regime. The use of accurate thermodynamic formulations capable of capturing the singular behavior of supercritical fluids allows for the efficiency and accuracy of each turbulence model to be tested, and their structure or the presence of specific terms linked to the characteristics of the final results.

A supercritical fluid is characterized by pressure and temperature above the fluid's critical values. In fuel injection phenomena, specifically in combustion chambers, both fuels and oxidizers' operating conditions can exceed their critical point as a means to increase the engine's efficiency [1–3]. At an arbitrary constant temperature, a gas can be converted to a liquid by increasing the pressure. As temperature increases, so does the kinetic energy of the molecules, requiring a higher pressure to bring the gas to a liquid. The critical temperature, T_c, marks the point after which a transition to the liquid phase is no longer possible, no matter the applied pressure. The vapor pressure at the critical temperature is then defined as the critical pressure, p_c. The critical point then marks the end of the vapor pressure line, where both temperature and pressure reach their critical values.

Since vaporization no longer occurs, a more suitable terminology is needed. Several authors, including [4], propose the use of "emission rate" and "emission constant" to describe mixing under supercritical conditions.

As the temperature increases even further, the liquid-like supercritical oxidizer crosses the pseudocritical line and transitions to a gas-like fluid. This transition from a liquid-like to a gas-like state could be compared to a subcritical boiling, the main difference being that the isothermal vaporization typical of subcritical fluids is replaced by a continuous non-equilibrium process that takes place over a finite temperature range (Figure 1). As this happens, the specific heat capacity goes through a maximum and tends to infinity when approaching the critical point, as shown in Figure 2. Similarly to [5], we refer to this transition phenomenon as pseudo-boiling and the maxima of the specific heat capacity as pseudo-boiling of pseudocritical line. It represents then a continuation of the saturation line well into the supercritical regime.

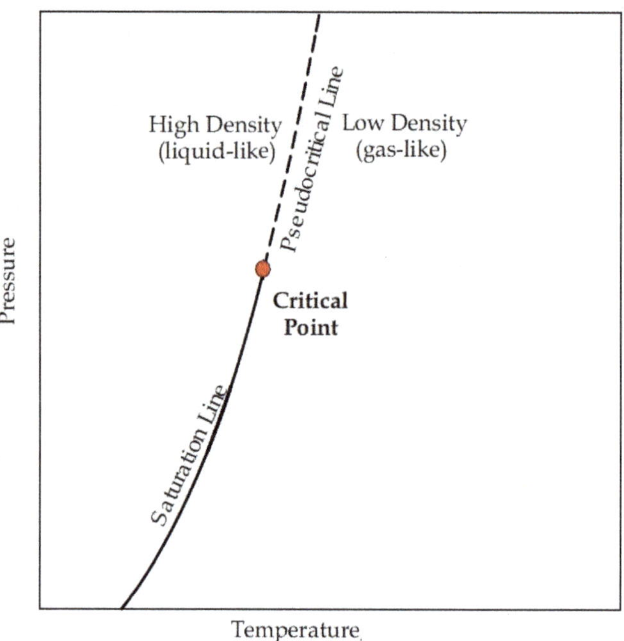

Figure 1. Overall pressure temperature diagram.

The two parameters $(\partial \rho / \partial T)$ and $c_{p_{max}}$ can therefore be used to identify pseudo-boiling temperatures for different pressure values as shown in Figure 2. For nitrogen, it is visible that as the pressure approximates the critical value of 3.39 MPa, the peak in specific heat becomes more noticeable along with the slope of $(\partial \rho / \partial T)$.

It is well known that in a subcritical injection, surface instabilities are responsible for jet atomization, small discrete ligaments begin to break up, and droplets are ejected from the jet core [6]. In a supercritical injection, however, the breakup mechanics are entirely different. [7] describes one of the main characteristics of supercritical fluids as the impossibility of a two-phase flow. Similar effects are reported by [5], where the surface tension is measured for oxygen from subcritical temperatures, with higher values, up to the critical temperature, for which it completely vanishes. Several other authors describe this different breakup mechanism where the drops and ligaments are no longer detected, and no distinct surface interface can be determined. [7] notes that this disintegration mechanism more closely resembles turbulent and diffusive mixing than the traditional jet disintegration and [8] describes a thermal-breakup mechanism where the limit of the jet core is defined by the transition of the fluid across the pseudocritical line.

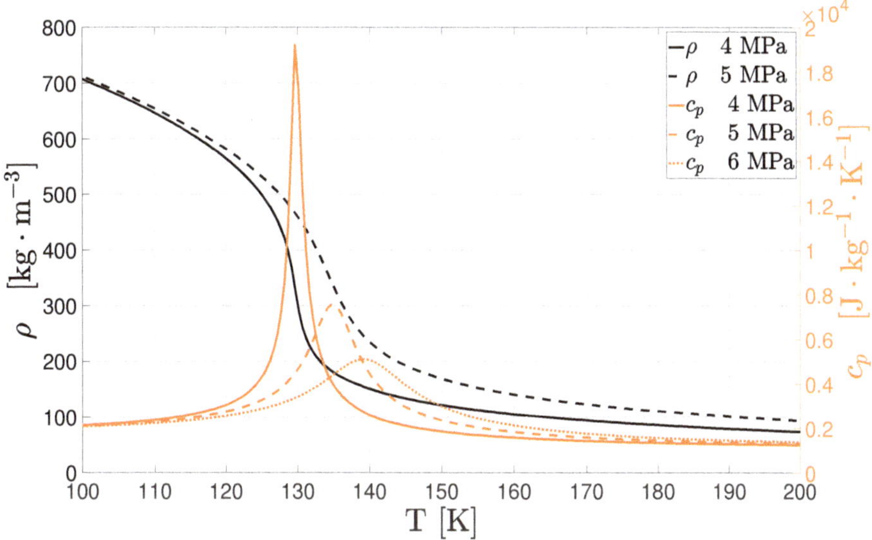

Figure 2. Density and isobaric specific heat values for nitrogen (data from the NIST database).

In the end, both the thermodynamic behavior and the breakup mechanisms have a direct effect on the jet structure. As the liquid-like nitrogen is injected into the chamber, its temperature increases as it begins to mix with a warmer gas, such as nitrogen. The structure of the flow changes and it can be divided into three characteristic regions: potential core, transition, and fully developed region. [9] defines the length of the potential core as the distance at which the centreline density remains relatively constant and [8] compiles and compares four different equations attempting to predict this length that is either based on the ratio between the densities of the liquid and gas-like fluids or are given a constant value for any specific test geometry. In the self-similar region, the absolute value of flow variables can still change, but their radial profiles are no longer a function of axial direction. In between these two regions lies the transition zone, where the turbulent and diffuse mixing is most relevant. As instabilities begin to appear, dense pockets of liquid-like nitrogen are separated from the jet core, causing an increase in density fluctuations [10,11]. As a result, the density sharply decreases, and the energy dissipation is significant. In experimental studies, this structure is visible through the axial and radial density distributions and also through the jet spreading angle.

Turbulence is regarded as the last unsolved problem of classical physics. The presence of advective terms in the governing equations leads to their admittance of a chaotic solution after a critical Reynolds number. Modeling is then performed resorting to techniques such as RANS (Reynolds Averaged Navier Stokes), LES (Large Eddy Simulation), or DNS (Direct Numerical Simulation). However, when discussing supercritical fluid flows, the fact that no turbulence models developed explicitly for flows at these conditions exist, is an added factor of uncertainty. Throughout the years, these techniques have been used in the modeling of supercritical fluid flows with various degrees of success.

In a series of studies by the same authors, [12–14], a Large Eddy Simulation solver is used for the computation of nitrogen injection. In this sense, [15] also perform LES following a pressure-based solution approach in which the PISO - Pressure-Implicit with Splitting of Operators algorithm for pressure-velocity coupling is employed. The authors conclude that further improvements are needed in the variables discretization so that the numerical diffusivity can be lowered.

The paths available in terms of discretization of the governing equations as well as in turbulence modeling are so diverse it justifies a study on how different approaches and modeling techniques affect the result. On this subject, [16] performs a comparison between different RANS derivations of the $\kappa - \epsilon$ model and an LES computation. Cubic equations of state performance are compared and coupled with the compressible formulation for the conservation of mass and momentum. As for the temperature and transport properties, they are retrieved as a function of the mixture fraction.

In [17], LES simulations are carried out, and a comparison is made between a density-based and a pressure-based solution. In the density-based solution, implicit LES combines turbulence modeling with the numerical discretization of the conservation equations, while in the pressure-based solution, an eddy viscosity approach is favored. The results comparison from the two solution approaches does not indicate a very substantial difference in the axial centerline distribution. Another LES pressure-based solution approach is proposed by [18], using different sub-grid scale models. Also, on the subject of supercritical injection, [19], conclude pressure-based solvers are strongly affected by the deviation of compressibility coefficients from the ideal gas behavior, and [20] propose an extension of a double-flux model to real fluid equations of state, as a means to improve the capability of the numerical solver to cope with spurious pressure oscillations, especially close to the pseudocritical line.

In terms of Direct Numerical Simulation (DNS), the work of [10] stands out, in which a lower inlet velocity is used, reducing the Reynolds number. A merged PISO/ SIMPLE algorithm allows for the study of entropy production rates, dominated by heat transfer [21].

In high Reynolds numbers simulations, the choice then lies between RANS- and LES-based simulations. While RANS relies on the same turbulence model for the entire inertial scales, in LES, the larger inertial scales are solved, and an SGS (Sub-Grid Scale) model is used for the smaller scales. A filter is used to separate resolvable and Sub-Grid Scales. The importance of Sub-Grid Scale modeling in high-pressure LES computations is outlined by [22]. Even though the potential of LES simulations is recognized, it has failed to outperform RANS-based solvers systematically. As a result, in the present work, a RANS-based numerical solver is used to compare the performance and accuracy of several turbulence models in the modeling of supercritical fluid flows. Specifically, the validation of a numerical setup replicating combustion chamber conditions can be seen as a first step towards accurately and reliably simulating combustion phenomena inside a Liquid Rocket Engine combustion chamber, and a small step in this direction is to understand how current turbulence models behave under supercritical conditions and how their accuracy stands up against one another. The same issue was dealt with in a previous work [23], through the reassessment of the concept of a variable turbulent Prandtl number, applied to the standard $\kappa - \varepsilon$ turbulence model.

Interestingly enough, the comparison described can be linked to the general state of the space sector. The dawn of space exploration privatizing has lead to a boost and renewed interest in the modeling of vehicles used for such missions. As a consequence, an ever-increasing demand for economic sustainability has lead to the necessity of developing new solutions for space technology. As stated by by [15], the new challenge is not the development of new technologies, but the improvement of already available concepts, with restricted budgets and shortened development cycles. In this sense, through the analysis of the RANS-based turbulence models here proposed an attempt is made to improve the knowledge regarding their behavior in the supercritical regime.

The remainder of this study is structured as follows: first, the experimental conditions necessary for validation purposes are shortly reviewed, and the initial and boundary conditions are established. The description of the governing equations in their Favre averaged formulation precedes the discussion about the adopted turbulence modeling; however, the models themselves are not detailed explained. Such an undertaking would not contribute to quality improvement of this manuscript, while would render it prohibitively extensive. A simple discussion of the models' limitations and applicability is favored. The processes of density and transport properties determination are then discussed, as well as the discretization of the governing equations. The results are then critically analyzed, and the conclusions reiterated for future studies.

2. Initial and Boundary Conditions

The test cases from [9] are the basis of comparison for this numerical study. In particular, case 3 and case 4 are investigated where cold nitrogen is injected into a chamber at ambient temperature with four windows for optical access.

The geometry corresponding to the experimental setup of [9] is represented in Figure 3. The diameter of the chamber and the injector are 122 mm and 2.2 mm, respectively, while measuring, in length, 250 mm and 90 mm. Liquid nitrogen is injected into a chamber filled with gaseous nitrogen according to the conditions indicated in Table 1.The subscript 0 respects to injection conditions, with ∞ representing conditions in the combustion chamber.

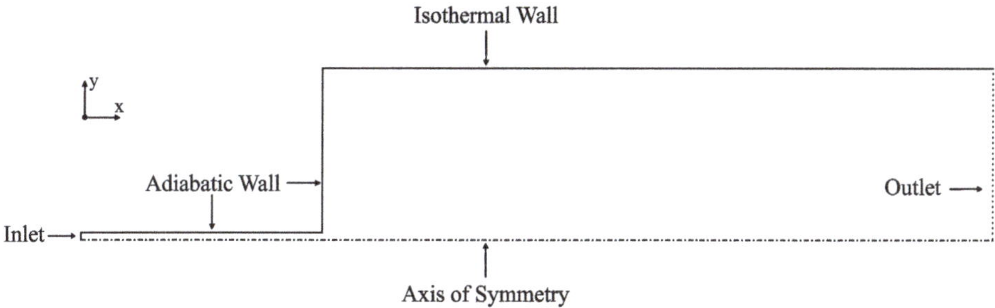

Figure 3. Boundary conditions for the physical model.

Table 1. Test conditions [9].

Case	p_∞ [MPa]	u_0 [m.s^{-1}]	T_0 [K]	T_∞ [K]	ρ_0 [kg.m^{-3}]	ρ_∞ [kg.m^{-3}]
3	3.97	4.9	126.9	297	457.82	45.24
4	3.98	5.4	137	297	164.37	45.36

The domain contains five different boundary conditions, also depicted in Figure 3. A constant axial velocity profile is set at the inlet to u_0, and the radial velocity is set to zero. At the walls, a no-slip condition is applied where both the normal and tangential velocity components are set to zero.

A pressure outlet is defined with a gauge pressure of 0 MPa and where the pressure values at the outlet face are calculated by averaging the specified operating pressure of p_∞, with the internal pressure. Also, at the symmetry axis, the value of any specific property is equaled to that of the adjacent cell.

Finally, for the adiabatic walls of the injector and the faceplate, the heat flux from Equation (1) is set to zero, but for the isothermal wall heat transfer is calculated through a Dirichlet boundary condition by setting a constant temperature at the wall of 297 K. In Equation (1), h_f represents the fluid heat transfer coefficient, T_w the temperature at the wall and T_∞ is the local fluid temperature.

$$q = h_f(T_w - T_\infty). \tag{1}$$

3. Governing Equations

To deal with the weakly incompressible but variable density conditions [24], the standard time-averaging method is replaced by the Favre averaging procedure, and the system of equations is closed with different turbulence models, the main focus of this study. The performance of said models, designed and calibrated to run in subcritical conditions, is then studied and their validity for the supercritical regime is assessed.

The Favre averaging method introduces a density-weighted quantity ($\tilde{\phi}_i$) and a density-weighted fluctuation (ϕ_i''), as given by equation (2) for an arbitrary scalar ϕ_i.

$$\phi_i = \tilde{\phi}_i + \phi_i''. \tag{2}$$

$\tilde{phi}_i$ is evaluated according to Equation (3), being $\bar{\rho}$ the mean density.

$$\tilde{\phi}_i = \frac{\overline{\rho\phi_i}}{\bar{\rho}}. \tag{3}$$

The steady-state Favre averaged conservation equations for mass, momentum, and energy are reproduced here, following the integral formulation in Equations (4) to (6), respectively, where $\tilde{u}$ represents density-weighted velocity components.

Mass can neither be created nor destroyed and as a result, there is no diffusive term in Equation (4), only an advective one.

$$\int_\Theta \left[\frac{\partial}{\partial x_i} (\bar{\rho}\tilde{u}_i) \right] d\Theta = 0. \tag{4}$$

Momentum is a vectorial quantity meaning that its transport is defined by as many equations as the number of dimensions that are assumed. The momentum advective flux is then defined in Equation (5) as $\bar{\rho}\tilde{u}_i\tilde{u}_j$, while $\bar{\rho}f_i$ represents the volume source term.

$$\int_\Theta \left[\frac{\partial}{\partial x_j} (\bar{\rho}\tilde{u}_i\tilde{u}_j) \right] d\Theta = \int_\Theta \left[-\frac{\partial \bar{p}}{\partial x_i} + \frac{\partial}{\partial x_j} \left(\tilde{t}_{ij} + \tau_{ij} \right) + \bar{\rho}f_i \right] d\Theta. \tag{5}$$

In Equation (6), the advective term is represented by $\bar{\rho}\tilde{u}_j\tilde{H}$, while the diffusive term is evaluated by Fourier's law of conduction, through q_j and τ_{ij} is the Reynolds stress tensor.

$$\int_\Theta \left[\frac{\partial}{\partial x_j} (\bar{\rho}\tilde{u}_j\tilde{H}) \right] d\Theta = \int_\Theta \left[\frac{\partial}{\partial x_j} \left[-q_j - q_{t_j} + \tilde{u}_i \left(\tilde{t}_{ij} + \tau_{ij} \right) + \overline{t_{ji}u_i''} - \overline{\tfrac{1}{2}\rho u_i'' u_i'' u_j''} \right] \right] d\Theta. \tag{6}$$

The viscous stress tensor from Equations (5) and (6) is averaged according to Equation (7), as a function of the molecular viscosiry where the mean strain-rate tensor, $\tilde{S}_{ij}$, is given by Equation (8) and δ_{ij} is the Kronecker delta function.

$$\tilde{t}_{ij} = 2\mu \left(\tilde{S}_{ij} - \frac{1}{3}\tilde{S}_{kk}\delta_{ij} \right) \tag{7}$$

$$\tilde{S}_{ij} = \frac{1}{2} \left(\frac{\partial \tilde{u}_i}{\partial x_j} + \frac{\partial \tilde{u}_j}{\partial x_i} \right). \tag{8}$$

The Reynolds stress tensor of Equation (5) holds correlations originated from the averaging process, given by Equation (9).

$$\tau_{ij} = -\overline{\rho u_i'' u_j''}. \tag{9}$$

Additional terms also appear in Equation (6) such as the Favre-averaged total specific energy ($\tilde{E}$), enthalpy ($\tilde{H}$) and the turbulent heat flux (q_{t_j}), defined in Equations (10), (11) and (12), respectively.

$$\tilde{E} = \tilde{e} + \frac{1}{2}\tilde{u}_i\tilde{u}_i + k \tag{10}$$

$$\tilde{H} = \tilde{h} + \frac{1}{2}\tilde{u}_i\tilde{u}_i + k \tag{11}$$

$$q_{t_j} = \overline{\rho u_j'' h''}. \tag{12}$$

Lastly, the turbulence kinetic energy per unit volume, k, is defined according to Equation (13).

$$\bar{\rho} k = \frac{1}{2}\overline{\rho u_i'' u_i''}. \tag{13}$$

The double and triple correlations from Equations (5) and (6) are also a product of the averaging process which is too extensive to describe here. While a physical meaning can be attributed to some, it does not apply to others. The Reynolds stress tensor alone introduces three additional independent variables, meaning that the system is not yet closed. A transport equation for τ_{ij} can be provided, but the number of unknowns only increases, and the system remains open. Nevertheless, this is the essence of second-order turbulence models. For now, an approximation for τ_{ij} is needed.

4. Turbulence Models

The Boussinesq approximation can be used to define the Reynolds stress tensor and is the basis of turbulence modeling. This approximation relates τ_{ij} with the viscous stress tensor by introducing the concept of eddy or turbulent viscosity, μ_t. It relates to the influence of molecular viscosity on the transport of momentum with the influence of turbulence viscosity on the transfer of momentum caused by turbulent fluctuations. As a result, Equation (7) becomes:

$$\tau_{ij} \approx 2\mu_t\left(\tilde{S}_{ij} - \frac{1}{3}\tilde{S}_{kk}\delta_{ij}\right) - \frac{2}{3}\bar{\rho}k\delta_{ij} \tag{14}$$

While the terms for $i \neq j$ are modeled through μ_t, the trace of τ_{ij} is still precisely defined through the specific turbulent kinetic energy from Equation (13) as:

$$\tau_{ii} = -\overline{\rho u_i'' u_i''} = -2\bar{\rho}k. \tag{15}$$

With this relation, turbulence models can focus on calculating the eddy viscosity (μ_t) and the turbulence kinetic energy (k).

The laminar and turbulent heat transport terms, q_j and q_{t_j}, are defined according to Fourier's law, so that:

$$q_j = -\frac{c_p\mu}{\mathrm{Pr}}\frac{\partial \tilde{T}}{\partial x_j} = -\frac{\mu}{\mathrm{Pr}}\frac{\partial \tilde{h}}{\partial x_j}, \qquad q_{t_j} = -\frac{c_p\mu_t}{\mathrm{Pr}_t}\frac{\partial \tilde{T}}{\partial x_j} = -\frac{\mu_t}{\mathrm{Pr}_t}\frac{\partial \tilde{h}}{\partial x_j}. \tag{16}$$

In Equation (16), Pr_t represents the turbulent Prandtl number, i.e., $\mathrm{Pr}_t = \nu_t/\alpha_t$, where ν_t is the eddy diffusivity of momentum and α_t the turbulent thermal diffusivity.

Finally, the molecular diffusion and the turbulent transport, $\overline{t_{ji}u_i''} - \overline{\rho\frac{1}{2}u_i''u_i''u_j''}$, are coupled together and modeled as shown in Equation (17).

$$\overline{t_{ji}u_i''} - \overline{\tfrac{1}{2}\rho u_i''u_i''u_j''} = \left(\mu + \frac{\mu_t}{\mathrm{Pr}_t}\right)\frac{\partial k}{\partial x_j}. \tag{17}$$

Turbulence models evaluate eddy viscosity in different ways but generally using the same properties, such as the turbulent kinetic energy, the turbulent dissipation rate, ε, and the specific dissipation rate, ω.

The mixing length hypothesis proposed by L. Prandtl provides an expression to define the turbulent viscosity based on the assumption that the x-momentum of fluid remains constant for a length of l_{mix} in the y-direction. l_{mix} is the mixing length that is characteristic of each flow geometry along with a characteristic velocity, that must be defined in advance. Ergo, zero equation models where the length and velocity scales are not defined through properties such as κ, ϵ or ω, are not independent of the case of study. One and two-equation models overcome this obstacle by introducing transport equations for history-dependent variables that can represent velocity and a length scale. Explicitly, the two-equation models studied in this work define the velocity scale through the turbulent kinetic energy to incorporate non-local and flow history effects in the turbulent viscosity. The underlying problem in RANS simulations results from the unavailability of velocity fluctuations, leading to the necessity of resorting to closure models. The chosen model involves, as do most of the choices made when attempting to numerically reproduce the behavior of supercritical fluid flows, a compromise between computational cost and accuracy. The current development of improved turbulence models faces the dilemma of conserving the low computational cost and high robustness of RANS approaches while incorporating as much physics as possible.

The Spalart–Allmaras model [25] is a one-equation model in which a direct derivation of an equation for the transport of the modified eddy viscosity is performed. This does not happen on the $k - \varepsilon$-based models [26–28], where transport equations for both the turbulent kinetic energy (κ) and its dissipation (ε) are introduced. In the $k - \omega$-based models, the dissipation of turbulent kinetic energy is replaced by the specific dissipation rate, and as such, a variation of the model of [29] is considered. The turbulent dissipation rate is replaced by the dissipation per unit time or specific dissipation rate defined as $\omega = \varepsilon/k$.

In the standard $\kappa - \omega$ model, additional terms related to low Reynolds numbers corrections and compressible effects are available for this model but are neglected for the current study. An attempt is made in reducing the round-jet anomaly by linking the dissipation of ω to the mean deformation of the flow. The predictions for k and ω outside of the shear layer remain on the most significant setbacks of this model, despite the introduction of a cross-diffusion term.

After the proposal of the Wilcox $k - \omega$ model, [30] developed the shear-stress transport (SST), $k - \omega$ model, to retain the robust and accurate formulation of the Wilcox model inside the shear layer while taking advantage of the free-stream qualities of the $k - \varepsilon$ model in the far-field. The transport equation for ε is converted into a similar formulation as that of ω. This blending function is designed to be one in the viscous sublayer of the boundary layer and tend to zero in the log-law region ($y^+ > 70$).

All the models reasoned insofar are based on Boussinesq's eddy viscosity concept. However, another one is considered, which does not have this concept as its underlying relationship. The Stress Baseline (BSL) model [30] closes the system of governing equations with transport equations for ω and τ_{ij}.

All these models are, however, subject to different levels of uncertainty, whose source identification remains a most challenging task due to the coupling between various phenomena and levels of uncertainty [31]. Nevertheless, several studies were conducted on uncertainty quantification in several turbulence models, and even though the test subjects are not supercritical fluid flows, their conclusions are broad and extensive.

The application of sensitivity analysis (defined as the derivatives of the flow variables concerning the design parameters) by [32] on a backward-facing step showed the calibration parameters C_1, and C_2, in the $\kappa - \epsilon$ turbulence model, have the most pronounced effect on the turbulence model predictions. On the Spalart–Allmaras model, [33] concludes the incomplete physical knowledge led to the use of dimensional analysis and a large amount of judgment to close or tune in model constants. Relations between coefficients need to be enforced to maintain the appropriate behavior of the Spalart–Allmaras model for canonical flows.

On the other hand, [34] presents a novel methodology for improving eddy viscosity models in predicting wall-bounded turbulent flows with strong gradients in the thermophysical properties. Conventional turbulence models for solving the RANS equations do not correctly account for variations in transport properties, such as density and viscosity.

A methodology in which representative samples of motions and processes of all scales are solved and combined, while remaining computationally affordable, especially at large Reynolds numbers, effectively meaning the statistical resolution of all scales remains a somewhat distant goal.

Additionally, the turbulent kinetic energy, turbulent dissipation rate, and specific dissipation rate are based on the turbulent intensity, I, and turbulent viscosity ratio, μ_t/μ:

$$k = \frac{3}{2}(Iu_0)^2, \quad \varepsilon = \rho C_\mu \frac{k^2}{\mu}\left(\frac{\mu_t}{\mu}\right)^{-1}, \quad \omega = \rho \frac{k}{\mu}\left(\frac{\mu_t}{\mu}\right)^{-1}. \tag{18}$$

In Equation (18), C_μ is specific to each of the turbulence models. In the Spalart–Allmaras model, $\tilde{\nu}$ is taken directly from the turbulent viscosity ratio, and in the Stress-BSL model, the turbulent stresses are assumed to be zero, except for the trace of the tensor which is calculated as in Equation (15). In all cases, the turbulence intensity is set to 5% at the inlet.

5. Equation of State and Transport Properties

In the supercritical regime, the ideal gas Equation of State (EoS) must be replaced by more accurate formulations. While multi-parameter EoS have high accuracy, they fall behind in terms of computational efficiency. To overcome this obstacle, in the present work, density values are loaded from a preexisting real gas library [35], based on a reference equation of state for nitrogen [36], allowing for increased accuracy at a reduced computational cost, since look-up tables are generated before the computations, effectively removing the need for thermophysical properties to be calculated in each iteration. The non-linearity of transport variables must also be accurately defined, and the equations proposed by [37] are used.

Multiparameter equations of state can be created through polynomial and exponential expansions where the coefficients multiplied by each term are specific to each fluid. These coefficients must be fitted through the available experimental data for the conditions in which the EoS is to be valid. The 32-term modified Benedict-Webb-Rubin (MBWR) [38] EoS achieves a relative density error smaller than 0.5% above and below the critical point, while [39] proposes a 12-term EoS with available coefficients for a series of substances, nitrogen included, while [40] provides a highly accurate 18-term EoS optimized directly for nitrogen.

The EoS presented by [36] is based on the Helmholtz energy, F, which is then normalized and set as a function of reduced temperature (τ) and density ($\delta = \rho/\rho_c$):

$$\frac{F(\rho, T)}{RT} = f(\delta, \tau) = f^0(\delta, \tau) + f^r(\delta, \tau). \tag{19}$$

The first right-hand-side term of equation (19) refers to the ideal gas contribution to the Helmholtz energy while the second represents the residual Helmholtz energy corresponding to the intermolecular forces considered in a real gas formulation.

The ideal gas contribution is defined in Equation (20), the residual addition is shown in Equation (21) and the corresponding constants are listed in [36]. Thermodynamic properties can then be calculated based on the derivatives of these two terms.

$$f^0(\delta, \tau) = \ln(\delta) + a_1\ln(\tau) + a_2 + a_3\tau + a_4\tau^{-1} + a_5\tau^{-2} + a_6\tau^{-3} + a_7\ln\left(1 - \exp[-a_8\tau]\right) \tag{20}$$

$$f^r(\delta, \tau) = \sum_{k=1}^{6} N_k \delta^{i_k} \tau^{j_k} + \sum_{k=7}^{32} N_k \delta^{i_k} \tau^{j_k} \exp[-\delta^{l_k}] + \sum_{k=33}^{36} N_k \delta^{i_k} \tau^{j_k} \exp[-\phi_k(\delta - 1)^2 - \beta_k(\tau - \gamma_k)^2]. \tag{21}$$

The coefficients specific to this EoS are obtained through data fitting methods based on experimental measurements from a series of authors and for a wide range of temperatures and pressures.

Transport properties have a direct impact on the governing equations. The substantial variation of transport properties as dynamic viscosity and thermal conductivity approaching and entering supercritical conditions is depicted for nitrogen in Figure 4.

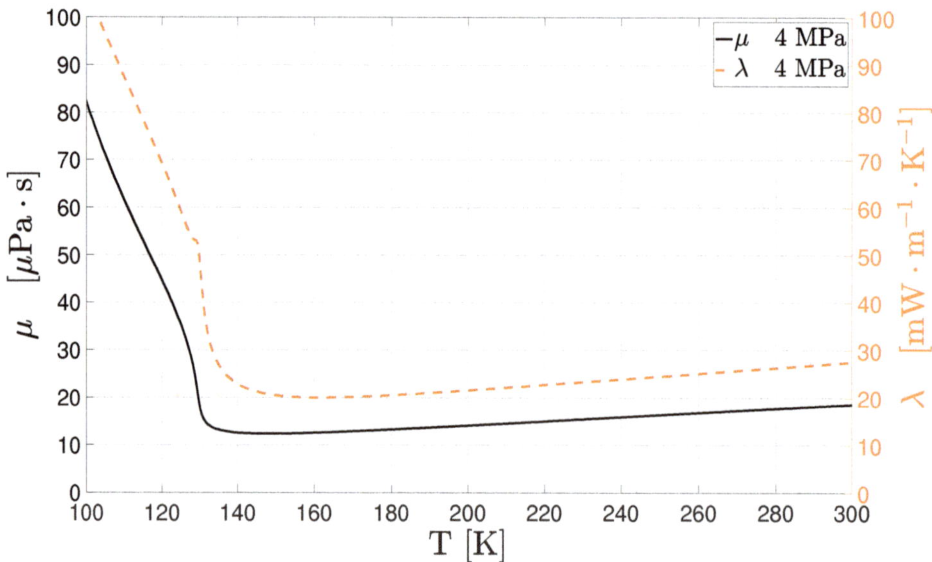

Figure 4. Viscosity and thermal conductivity values for nitrogen (data from the NIST database).

Following [37], viscosity is expressed according to Equation (22). Viscosity is evaluated from a dilute gas contribution, $\mu^0(T)$ (Equation (23)) and a residual component, $\mu^r(\tau, \delta)$, (Equation (24)). The σ represents the Lennard-Jones parameter and Ω the collision integral, while the remaining coefficients are tabulated in [37].

$$\mu = \mu^0(T) + \mu^r(\tau, \delta) \tag{22}$$

$$\mu^0(T) = \frac{0.0266958\sqrt{MT}}{\sigma^2 \Omega(T^*)} \tag{23}$$

$$\mu^r(\tau, \delta) = \sum_{i=1}^{n} N_i \tau^{t_i} \delta^{d_i} \exp(-\gamma \delta^{l_i}). \tag{24}$$

In the same fashion, thermal conductivity is defined according to Equation (25). The dilute and residual gas contributions are defined according to Equations (26) and (27), respectively.

$$\lambda = \lambda^0(T) + \lambda^r(\tau, \delta) + \lambda^c(\tau, \delta) \tag{25}$$

$$\lambda^0 = N_1 \left[\frac{\eta^0(T)}{1\,\text{Pa.s}} \right] + N_2 \tau^{t_2} + N_3 \tau^{t_3} \tag{26}$$

$$\lambda^r = \sum_{i=4}^{n} N_i \tau^{t_i} \delta^{d_i} \exp(-\gamma_i \delta^{l_i}). \tag{27}$$

A third component is needed for the evaluation of thermal conductivity in the critical region. Such contribution is defined in Equation (28). The calculation of the Ω_i coefficients comes from the definition of specific heat at constant pressure and volume.

$$\lambda^c = \rho C_p \frac{K R_0 T}{6\pi \xi \eta(T, \rho)} (\tilde{\Omega} - \tilde{\Omega}_0). \tag{28}$$

The use of real gas relationships for transport and thermodynamic properties allows the physical model to capture the weak compressibility effects when using an incompressible variable-density approach. Thermodynamic properties such as enthalpy are evaluated by their ideal gas value with a departure function to account for real gas effects.

6. Numerics

The values of the scalar ϕ are stored at the center of the cells. However, face values ϕ_r and ϕ_l are also necessary and must be interpolated from the cell-centered values. The diffusion of a certain quantity, as an example, is affected by the gradient of concentration of that same quantity over the entire domain, and a central difference is considered more appropriate. It is, therefore, used for the diffusive terms of the conservation equations.

On the other hand, as [16] suggests, at least a second-order upwind scheme is necessary for the modeling of the advective fluxes. The QUICK scheme [41] is employed instead as a tool for reducing the oscillatory and unstable behavior of second-order numerical schemes, while also dealing with the numerical diffusion affecting the first-order upwind schemes. Equation (29) serves as an example to demonstrate the concept. At point C, it can be discretized using the values of ϕ at the cell faces, as shown in equation (30), where Γ represents the diffusivity coefficient. These, however, need to be interpolated through the stored cell-centered values.

$$\frac{\partial \phi}{\partial t} = -\frac{\partial(u\phi)}{\partial x} + \frac{\partial}{\partial x}\left(\Gamma \frac{\partial \phi}{\partial x}\right) \tag{29}$$

$$\frac{\partial \phi_C}{\partial t} = \left[u_l \phi_l - u_r \phi_r + \Gamma_r \left(\frac{\partial \phi_r}{\partial x}\right) - \Gamma_l \left(\frac{\partial \phi_l}{\partial x}\right) \right] / \Delta x_C. \tag{30}$$

The application of a central differencing scheme to the diffusive term of Equation (30) has a stabilizing effect and is therefore straightforward. However, when applied to the advective term, it can lead to instabilities and an oscillatory behavior for a grid Péclet number (Pe) higher than two, i.e., local advection two times larger than diffusion. In short, the second-order accuracy can come at the expense of stability. By contrast, in an upwind differencing scheme, the cell-centered value of ϕ is assumed to represent a cell average value and hold throughout the entire cell, meaning that the face quantities are identical to the upstream cell quantities. This technique provides increased stability of the advective term to the variations of ϕ_C, but only because of the numerical diffusion introduced by assuming $\phi_l = \phi_L$. To diminish this numerical diffusion the grid spacing must considerably decrease, leading to a higher computational cost which is also not desirable.

The QUICK scheme combines a higher-order accuracy with the directional behavior of the upwind scheme to provide additional stability for the advective term in a coarser mesh. The face values are defined according to Equations (31) and (32). In this way the QUICK scheme achieves a third-order accuracy.

$$\phi_r = \frac{1}{2}(\phi_C + \phi_R) - \frac{1}{8}(\phi_L + \phi_R - 2\phi_C) \tag{31}$$

$$\phi_l = \frac{1}{2}(\phi_L + \phi_C) - \frac{1}{8}(\phi_{FL} + \phi_C - 2\phi_L). \tag{32}$$

A pressure-based algorithm is implemented where conservation of mass is implicitly achieved through a pressure-based continuity equation, obtained by taking the divergent of the momentum Equation (5) and introducing the condition $(\partial \bar{\rho} \tilde{u}_i)(\partial x_i) = 0$. A system of equations comprising this and the momentum equations is solved to obtain the velocity and pressure fields simultaneously. Energy and transport equations for turbulent variables are solved until convergence is reached.

In a collocated grid scheme, both the velocity and the pressure values needed for interpolation are retrieved from the same cell. However, when calculating the pressure field on a collocated grid, oscillations in the pressure field may appear as a result of an odd-even decoupling of the pressure and velocity, i.e., that on a specific point the pressure and velocity do not affect one another [42]. As a result, a staggered grid method [43] is used, where the velocity and pressure values are stored in different positions and for which the control volumes are no longer equal. Ultimately, the pressure values are calculated directly for the cell face, and no interpolation is needed. The decoupling of the pressure and velocity fields is eliminated along with any possible oscillations and is therefore used in the current work.

A hybrid initialization method is employed where the inlet velocity is set to u_0 from Table 1 and the absolute pressure at the outlet is set to p_∞. The velocity and pressure fields calculated with this method are then introduced in the first cycle of the pressure-based algorithm. A flowchart of the numerical procedure is given in Figure 5.

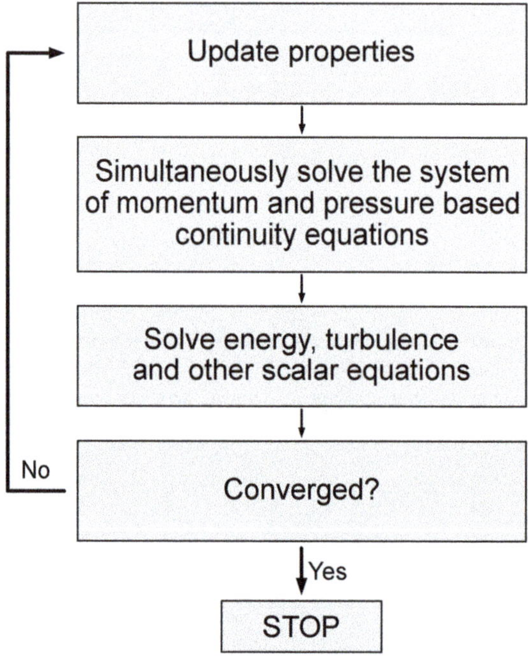

Figure 5. Pressure-based solution algorithm.

7. Grid Independence

An independence mesh study based on the centreline decay of the density is performed using three levels of refinement with 180 000, 280 000, and 495 000 points in a structured orthogonal mesh of rectangular elements. The comparison is made for case 4 from Table 1, and the results, with the standard $k - \varepsilon$ model, are shown in Figure 6. Despite a very slight variation of density values in the transition region, the three meshes provide close results to one another with similar slopes, indicating that the flow is sufficiently well resolved with the coarser grid. The more refined grid is not applied because the gained accuracy does not justify the additional computational cost, and the mesh of 280 000 points is used over the coarser one to maintain grid independence for the remaining turbulence models.

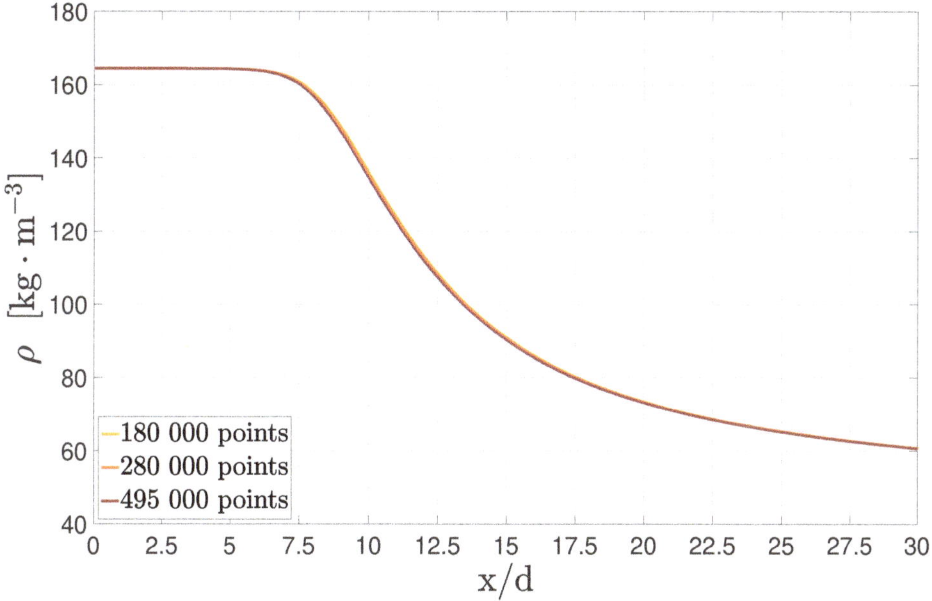

Figure 6. Centreline density decay at three different grid resolutions for case 4 with the standard $k - \varepsilon$ model and the REFPROPv9.1.

8. Results

Figure 7 shows a comparison of the results obtained for the centerline density decay when using the turbulence models described for case 3. It is visible that almost all models predict a potential core with values ranging between $x/d = 6.4$ and $x/d = 7.6$. The only exception is the standard $k - \omega$ model that largely overestimates the length of the potential core to $x/d \approx 12.5$, which can be attributed to the poor performance of this model in free-stream conditions. Even if the version here tested is an improvement over the 1998 Wilcox $k - \omega$ model, with an added cross-diffusion term introduced specifically to deal with the free-stream sensitivity, it does not provide acceptable results throughout the whole of the domain.

The density values predicted in this region are higher than those of the experimental data, but this can be a result of the measuring procedure used by [9]. Raman spectroscopy works by directing a monochromatic laser at the test substance and measuring the scattered radiation using a sensor. However, for higher densities, the jet tends to deflect the radiation along the axial direction, thus decreasing signal intensity at the sensor. As density reduces, this phenomenon is no longer predominant.

Turbulence seems not to influence the potential core. However, when entering the transition region, instabilities start to appear, and turbulent dissipation begins to increase. [10] reports a maximum of density fluctuations in this region as pockets of dense fluid start to smear the potential core. The same authors also discuss how the heat absorbed to overcome the intermolecular attraction leads to an increase in the heat entropy production with a maximum already closer to the self-similar region. An analogy can be found between this and the trend of κ and ϵ.

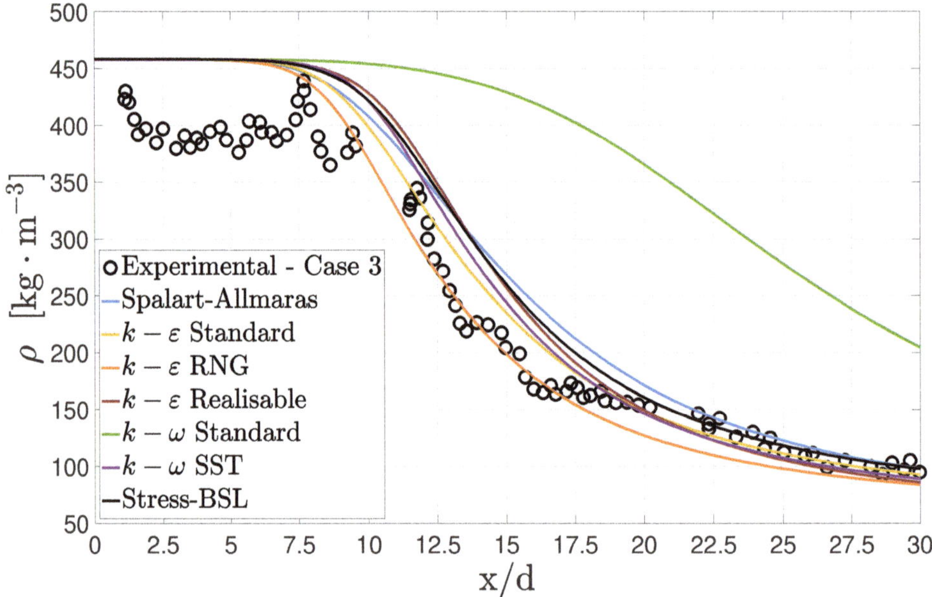

Figure 7. Centerline density decay in case 3 with different turbulence models.

When this happens, the fluid crosses the pseudocritical line leading to thermal expansion and reduced shearing, resulting in turbulence dissipation, as depicted in Figure 8, starting to decrease around $x/d = 15$. This thermal expansion also appears to spawn sharp velocity fluctuations visible in the increase of turbulent kinetic energy, indicative of a robust turbulent mixing mechanism. The comparison of the maxima of c_p along the axial density evolution is seen in Figure 9 for the RNG $\kappa - \epsilon$ model.

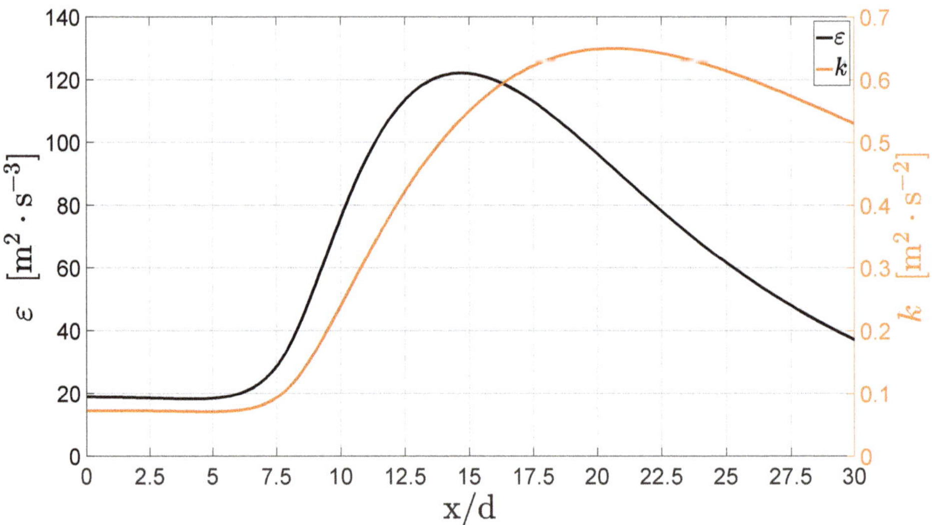

Figure 8. Centreline distribution of the turbulent dissipation rate and kinetic energy in case 3 with the standard $k - \epsilon$ model.

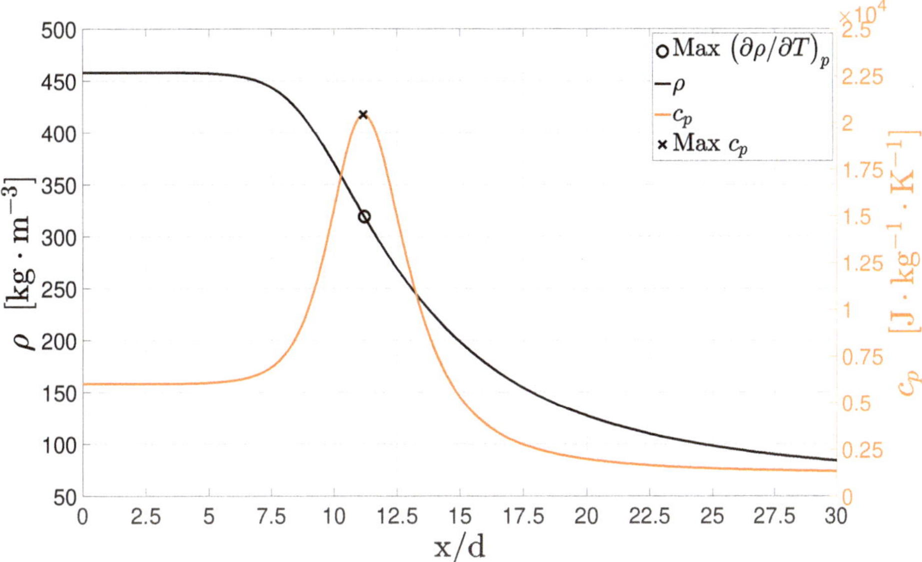

Figure 9. Location of the maxima of c_p in the axial density decay for case 3 with the RNG $\kappa - \epsilon$ turbulence model.

In this region, the more elaborate structure of the SST $k - \omega$ does not provide outstanding results despite having a shear-stress-based formulation for eddy viscosity. When moving away from the wall, μ_t returns to $\overline{\rho}k/\omega$, and the model only uses the $k - \varepsilon$-based coefficients. Consequently, we can see that its results more closely match those of the $k - \varepsilon$ variants than those of the standard $k - \omega$ model.

The five equation Stress-BSL model also overestimates the density values between $7.5 < x/d < 20$, which could be related, in part, to the dependency of this model on ω and the inherent deficiencies of its transport equation. The one equation Spalart–Allmaras leads to a similar overestimation of density values between $7.5 < x/d < 22.5$, but it is striking to see how well this simple formulation stacks against a five equation model.

In the three ε based models, the realizable variant provides the worst results since the beginning of the transition zone up to $x/d \approx 20$. It seems that the realisability constraint and the alternative ε transport equation do not give any visible contribution. Between the standard and the RNG $k - \varepsilon$, there are two main differences: the formulation for the destruction of the turbulent dissipation rate and the definition for the turbulent Prandtl number inserted into the energy equation and in the turbulent variables transport equations. Especially when considering the work from [23], we are led to believe that the variable turbulent Prandtl number is the leading cause for an improved behavior of the RNG $k - \epsilon$ model over the standard version. The model accurately predicts the density values across the domain.

In case 4, the differences in potential core length, depicted in Figure 10, are similar to those of case 3, ranging from $x/d = 6.6$ to $x/d = 7.7$ with the standard $k - \omega$ being once again the exception. Injection density is considerably reduced in this case when compared to that of case 3, and, as a consequence, there is no longer an apparent density overestimation in the potential core. However, an unrealistic potential core is still predicted independently of the turbulence model. In this case, the Stress-BSL model is the only one to correctly predict density values between $17.5 < x/d < 30$ while the remaining provide a slight under prediction. Nevertheless, its behavior outside this region is not exceptional, and the overall results do not justify the additional computational cost. Results from the RNG $k - \epsilon$ continue to be acceptable, but there is also an improvement from the standard $k - \varepsilon$ and the Spalart–Allmaras models for which the outcome is nearly identical. The decrease in the turbulence energy dissipation happens soon, at an $x/d = 12.5$ (Figure 11) As it can be seen in Figure 12, since the pseudo-boiling line is not crossed, there is not a peak in the c_p, which has an evolution in accordance to the axial density decay. The injection temperature of 137 K is already above the pseudo-boiling temperature, and there is no peak in the specific heat. remaining fairly constant until $x/d \approx 5$ along with the temperature inside the potential core, and begins then to decrease.

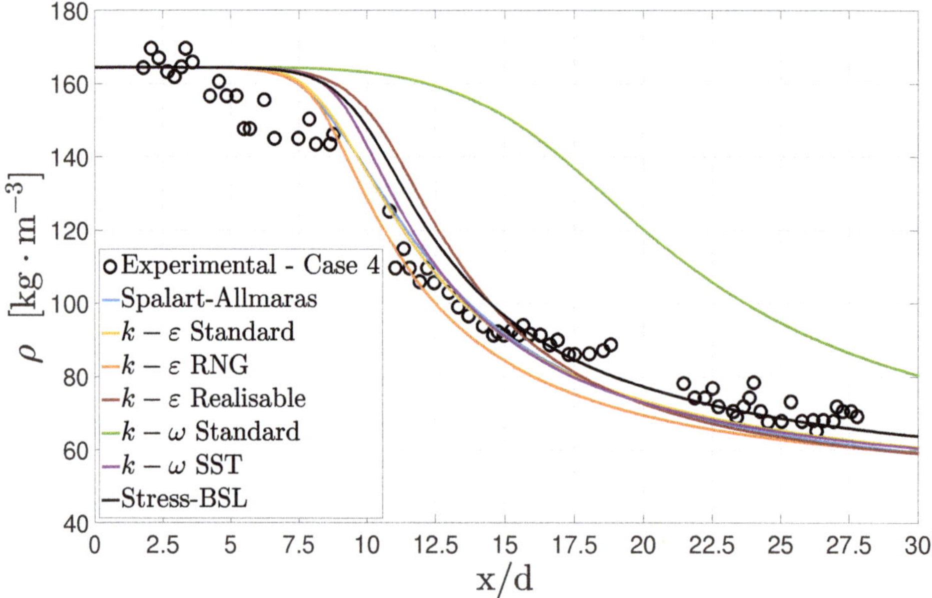

Figure 10. Centreline density decay in case 4 with different turbulence models.

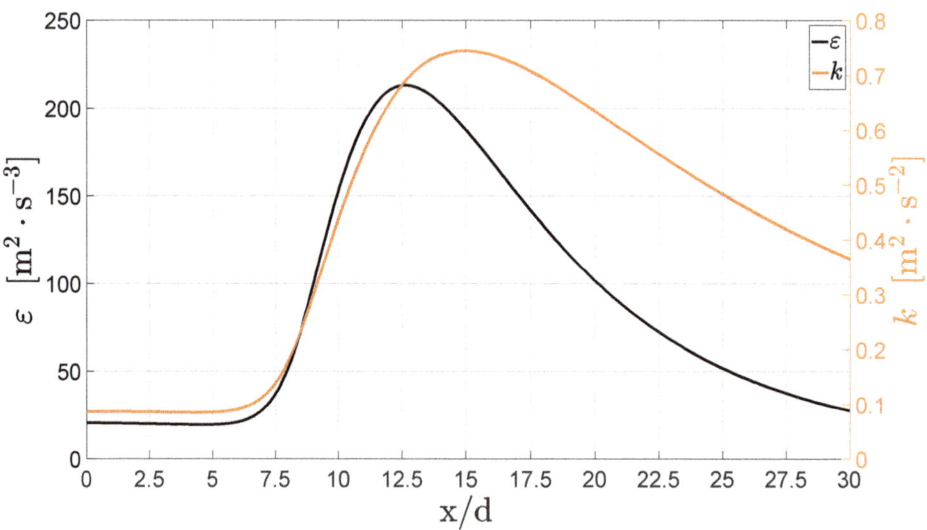

Figure 11. Centreline distribution of the turbulent dissipation rate and kinetic energy in case 4 with the RNG $k - \epsilon$ mode.

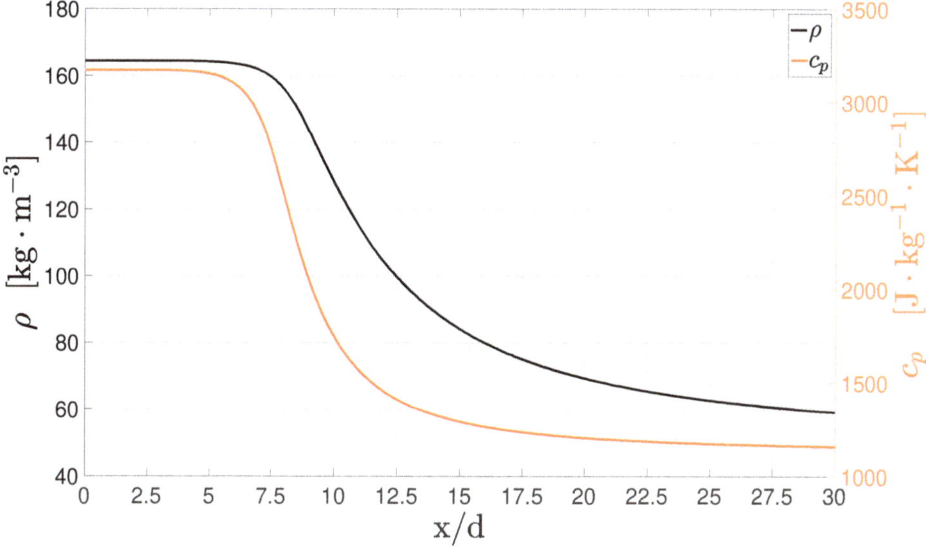

Figure 12. Location of the maxima of c_p in the axial density decay for case 4 with the RNG $\kappa - \epsilon$ turbulence model.

9. Conclusions

The steady-state Favre averaged governing equations are used to deal with the incompressible but variable-density flow that is characteristic of the current test cases. The system of equations is closed with six different models that are based either on the turbulent viscosity or the transport of the Reynolds stresses, to study the behavior of turbulence modeling in supercritical conditions.

An accurate formulation, capable of replicating the singular behavior of transport properties of supercritical fluids, is reviewed with detail. A pressure-based algorithm is used where the velocity and pressure fields are solved simultaneously. A staggered grid method is then implemented to prevent pressure fluctuations together with the QUICK scheme for the advective terms and second-order central differencing scheme for the diffusive terms.

The results obtained are compared to the experimental data for validation. There is a generally good agreement with the experimental data for both case 3 with the RNG $k - \varepsilon$ model and for case 4 also with the RNG $k - \varepsilon$ and with the Spalart–Allmaras model. Nevertheless, there is a clear distinction is that the results obtained from different turbulence models.

The results obtained with a second-order turbulence model show that there is no clear advantage in calculating higher-order turbulent correlation terms, indicating that their relevance under these conditions is minimal. Also, the blending functions and shear tress transport formulation for μ_t of the SST $k - \omega$ model do not contribute to improving predictions since the flow is mainly in a free-stream region. The RNG $k - \varepsilon$ model, offers the best results for case 3, possibly due to the variable turbulent Prandtl number but the similarly good results obtained for case 4 with the Spalart–Allmaras and the Standard $k - \varepsilon$ models indicate that there may be other relevant factors.

As a consequence, the results here obtained are indicative that there is no direct correlation between the turbulence model complexity and quality of the results, in what supercritical fluid flows are concerned. These results indicate that computational time can be gained with the use of simpler turbulence models. For this fact also contributes the pre-compiled real gas equation of state.

Having determined how the velocity field affects the flow structure, one of the next step is, therefore, to determine the impact of a heating mechanism inside the injector and the correct boundary conditions to be used.

Energies **2020**, *13*, 1586

Author Contributions: Formal analysis, L.M. and F.C.; Funding acquisition, A.S. and J.B.; Supervision, A.S. and J.B.; Validation, F.C.; Writing—original draft, L.M.; Writing—review & editing, L.M. and A.S. All authors have read and agreed to the published version of the manuscript.

Funding: The present work was performed under the scope of the Aeronautics and Astronautics Research Center (AEROG) of the Laboratório Associado em Energia, Transportes e Aeronáutica (LAETA) activities and it was supported by Fundação para a Ciência e Tecnologia (FCT) through grant number SFRH/BD/136381/2018 and project numbers UID/EMS/50022/2019 and UIDB/50022/2020.

Conflicts of Interest: The authors declare no conflict of interest. The funders had no role in the design of the study; in the collection, analyses, or interpretation of data; in the writing of the manuscript, or in the decision to publish the results.

References

1. Brunner, G. Applications of supercritical fluids. *Annu. Rev. Chem. Biomol. Eng.* **2010**, *1*, 321–342. doi:10.1146/annurev-chembioeng-073009-101311. [CrossRef]

2. Haeseler, D.; Haidinger, F.; Brummer, L.; Haberle, J.; Luger, P. Development and testing status of the Vinci thrust chamber. In Proceedings of the 48th AIAA/ASME/SAE/ASEE Joint Propulsion Conference & Exhibit, Atlanta, GA, USA, 30 July–1 August 2012, doi:10.2514/6.2012-4334. [CrossRef]

3. Lacaze, G.; Misdariis, A.; Ruiz, A.; Oefelein, J. Analysis of high-pressure Diesel fuel injection processes using LES with real-fluid thermodynamics and transport. *Proc. Combust. Inst.* **2015**, *35*, 1603–1611, doi:10.1016/j.proci.2014.06.072. [CrossRef]

4. DeSouza, S.; Segal, C. Sub- and supercritical jet disintegration. *Phys. Fluids* **2017**, *29*, 47–107, doi:10.1063/1.4979486. [CrossRef]

5. Oschwald, M.; Smith, J.J.; Braman, R.; Hussong, J.; Schik, A.; Chehroudi, B.; Talley, D.G. Injection of fluids into supercritical environments. *Combust. Sci. Technol.* **2006**, *178*, 49–100, doi:10.1080/00102200500292464. [CrossRef]

6. Mayer, W. Coaxial atomization of a round liquid jet in a high speed gas stream: A phenomenological study. *Exp. Fluids* **1994**, *16*, 401–410, doi:10.1007/BF00202065. [CrossRef]

7. Bellan, J. Supercritical (and subcritical) fluid behavior and modeling: drops, streams, shear and mixing layers, jets and sprays. *Prog. Energy Combust. Sci.* **2000**, *26*, 329–366, doi:10.1016/S0360-1285(00)00008-3. [CrossRef]

8. Banuti, D.T. Crossing the Widom-line – supercritical pseudo-boiling. *J. Supercrit. Fluids* **2015**, *98*, 12–16, doi:10.1016/j.supflu.2014.12.019 ISSN 0896-8446. [CrossRef]

9. Mayer, W.; Telaar, J.; Branam, R.; Schneider, G.; Hussong, J. Raman measurements of cryogenic injection at supercritical pressure. *Heat Mass Transf.* **2003**, *39*, 709–719, doi:10.1007/s00231-002-0315-x. [CrossRef]

10. Ries, F.; Janicka, J.; Sadiki, A. Thermal transport and entropy production mechanisms in a turbulent round jet at supercritical thermodynamic conditions. *Entropy* **2017**, *19*, 404, doi:10.3390/e19080404. [CrossRef]

11. Ries, F.; Obando, P.; Shevchuck, I.; Janicka, J.; Sadiki, A. Numerical analysis of turbulent flow dynamics and heat transport in a round jet at supercritical conditions. *Int. J. Heat Fluid FLow* **2017**, *66*, 172–184, doi:10.1016/j.ijheatfluidflow.2017.06.007. [CrossRef]

12. Schmitt, T.; Selle, L.; Cuenot, B.; Poinsot, T. Large-Eddy Simulation of transcritical flows. *Comptes Rendus Mécanique* **2009**, *337*, 528–538, doi:10.1016/j.crme.2009.06.022. [CrossRef]

13. Schmitt, T.; Selle, L.; Ruiz, A.; Cuenot, B. Large-Eddy Simulation of supercritical-pressure round jets. *AIAA J.* **2010**, *48*, 2133–2144, doi:10.2514/1.J050288. [CrossRef]

14. Schmitt, T.; Ruiz, A.; Selle, L.; Cuenot, B. Numerical investigation of destabilization of supercritical round turbulent jets using Large Eddy Simulation. *Prog. Propuls. Phys.* **2011**, *2*, 225–238, doi:10.1051/eucass/201102225. [CrossRef]

15. Jarczyk, M.; Pfitzner, M. Large Eddy Simulation of supercritical nitrogen jets. In *50th AIAA Aerospace Sciences Meeting including the New Horizons Forum and Aerospace Exposition*; AIAA: Nashville, TN, USA, doi:10.2514/6.2012-1270. [CrossRef]

16. Park, T.S. LES and RANS simulations of cryogenic liquid nitrogen jets. *J. Supercrit. Fluids* **2012**, *72*, 232–247, doi:10.1016/j.supflu.2012.09.004. [CrossRef]

17. Müller, H.; Niedermeier, C.A.; Jarczyk, M.; Pfitzner, M.; Hickel, S.; Adams, N.A. Large Eddy Simulation of trans- and supercritical injection. *Prog. Propuls. Phys.* **2016**, *8*, 5–24, doi:10.1051/eucass/201608005. [CrossRef]

18. Taghizadeh, S.; Jarrahbashi, D. Proper orthogonal decomposition analysis of turbulent cryogenic liquid jet injection under transcritical and supercritical conditions. *Atom. Sprays* **2018**, *28*, 875–900, doi:10.1615/AtomizSpr.2018028999. [CrossRef]

19. Li, L.; Xie, M.; Wei, W.; Jia, M.; Liu, H. Numerical investigation on cryogenic liquid jet under transcritical and supercritical conditions. *Cryogenics* **2018**, *89*, 16–28, doi:10.1016/j.cryogenics.2017.10.021. [CrossRef]

20. Ma, P.C.; Lv, Y.; Ihme, M. An entropy-stable hybrid scheme for simulations of transcritical real-fluid flows. *J. Comput. Phys.* **2017**, *340*, 330–357, doi:10.1016/j.jcp.2017.03.022. [CrossRef]

21. Sierra-Pallares, J.; del Valle, J.G.; Garcia-Carrascal, P.; Ruiz, F.C. Numerical study of supercritical and tanscritical injection using different turbulent Prandtl numbers. A second law analysis. *J. Supercrit. Fluids* **2016**, *115*, 86–98, doi:10.1016/j.supflu.2016.05.001. [CrossRef]

22. Gnanaskandan, A.; Bellan, J. Large Eddy Simulations of high pressure jets: Effect of subgrid scale modeling. In *55th AIAA Aerospace Sciences Meeting*; AIAA: Reston, VA, USA, 2017, doi:10.2514/6.2017-1105. [CrossRef]

23. Magalhães, L.; Silva, A.; Barata, J. Locally variable turbulent Prandtl number considerations on the modeling of liquid rocket engines operating above the critical point. In Proceesings of the ILASS Europe 2019—29th European Conference on Liquid Atomization and Sprays, Paris, France, 2–4 September 2019. [CrossRef]

24. Barata, J.; Gökalp, I.; Silva, A. Numerical study of cryogenic jets under supercritical conditions. *J. Propul. Power* **2003**, *19*, 142–147, doi:10.2514/2.6090.

25. Spalart, P.; Allmaras, S. A one-equation turbulence model for aerodynamic flows. In Proceedings of the 30th Aerospace Sciences Meeting and Exhibit, Reno, NV, USA, 6–9 January 1992, doi:10.2514/6.1992-439. [CrossRef]

26. Launder, B.E.; Spalding, D.B. *Lectures in Mathematical Models of Turbulence*; Academic Press: London, UK, 1972. [CrossRef]

27. Yakhot, V.; Orszag, S.A.; Thangam, S.; Gatski, T.B.; Speziale, C.G. Developmentt of turbulence models for shear flows by a double expansion technique. *Physi. Fluids A Fluid Dyn.* **1992**, *4*, 1510–1520, doi:10.1063/1.858424. [CrossRef]

28. Yang, Z.; Shih, T.H. New time scale based k-epsilon model for near-wall turbulence. *AIAA J.* **1993**, *31*, 1191–1198. [CrossRef]

29. Wilcox, D.C. *Turbulence Modeling for CFD*, 2nd ed.; DCW Industries: Palm Dr La Canada, CA, USA, 1998.

30. Menter, F.R. Two-equation eddy-viscosity turbulence models fo engineering applications. *AIAA J.* **1994**, *32*, 1598–1605, doi:10.2514/3.12149. [CrossRef]

31. Xiao, H.; Cinnella, P. Quantification of model uncertainty in RANS simulations: a review. *Prog. Aerosp. Sci.* **2019**, *108*, 1–31, doi:10.1016/j.paerosci.2018.10.001. [CrossRef]

32. Turgeon, É.; Pelletier, D.; Borggaard, J.; Etienne, S. Application of a sensitivity equation method to the k–ε model of turbulence. *Optim. Eng.* **2007**, *8*, 341–372, doi:10.1007/s11081-007-9003-5.

33. Schaefer, J.A.; Cary, A.W.; Mani, M.; Spalart, P.R. Uncertainty quantification and sensitivity analysis of SA turbulence model coefficients in two and three dimensions. In *55th AIAA Aerospace Sciences Meeting*; AIAA: Reston, VA, USA, 2017, doi:10.2514/6.2017-1710. [CrossRef]

34. Otero, G.J.; Patel, A.; Diez, R.; Pecnik, R. Turbulence modelling for flows with strong variations in thermo-physical properties. *Int. J. Heat Fluid Flow* **2018**, *73*, 114–123, doi:10.1016/j.ijheatfluidflow.2018.07.005. [CrossRef]

35. Lemmon, E.W.; Bell, I.H.; Huber, M.L.; McLinden, M.O. *NIST Standard Reference Database 23: Reference Fluid Thermodynamic and Transport Properties-REFPROP, Version 10.0*; National Institute of Standards and Technology: Gaithersburg, MD, USA, 2018. [CrossRef]

36. Span, R.; Lemmon, E.W.; Jacobsen, R.T.; Wagner, W.; Yokozeki, A. A reference equation of state for the thermodynamic properties of nitrogen for temperatures from 63.151 to 1000 K and pressures to 2200 MPa. *J. Phys. Chem. Refe. Data* **2000**, *29*, 1361–1433, doi:10.1063/1.1349047. [CrossRef]

37. Lemmon, E.W.; Jacobsen, R.T. Viscosity and thermal conductivity equations for nitrogen, oxygen, argon and air. *Int. J. Thermophys.* **2004**, *25*, 21–69, doi:10.1023/B:IJOT.0000022327.04529.f3. [CrossRef]

38. Younglove, B.A. Thermophysical properties of fluids. I. argon, ethylene, parahydrogen, nitrogen, nitrogen trifluoride and oxygen. *J. Phys. Chem. Ref. Data* **1982**, *11*, doi:10.1063/1.555731.

39. Span, R.; Wagner, W. Equations of state for technical applications. I. Simultaneously optimized functional forms for nonpolar and polar fluids. *Int. J. Thermophys.* **2003**, *24*, 1–39. [CrossRef]
40. Span, R.; Lemmon, E.W.; Jacobsen, R.T.; Wagner, W. A reference quality equation of state for nitrogen. *Int. J. Thermophys.* **1998**, *19*, 1121–1132, doi:10.1023/A:1022689625833. [CrossRef]
41. Leonard, B.P. A stable and accurate convective modelling procedure based on quadratic upstream interpolation. In *Computer Methods in Applied Mechanics and Enginnering*; Elsevier: Amsterdam, The Netherlands, 1979; pp. 58–98. [CrossRef]
42. Hirsh, C. *Numerical Computation of Internal & External Flows*, 2nd ed.; Butterworth-Heinemann: Oxford, UK, 2007; Volume 1. [CrossRef]
43. Patankar, S.V. *Numerical Heat Transfer and Fluid Flow*; series in computational methods in mechanics and thermal sciences; Hemisphere Publishing Company: Milton Park, UK, 1972. [CrossRef]

Article

Numerical Investigation of Frequency and Amplitude Influence on a Plunging NACA0012

Emanuel Camacho, Fernando Neves, André Silva * and Jorge Barata

LAETA-Aeronautics and Astronautics Research Center, University of Beira Interior, 6201-001 Covilhã, Portugal; emanuel.camacho@ubi.pt (E.C.); fernandomneves@gmail.com (F.N.); jbarata@ubi.pt (J.B.)
* Correspondence: andre@ubi.pt

Received: 27 February 2020; Accepted: 4 April 2020; Published: 11 April 2020

Abstract: Natural flight has always been the source of imagination for Mankind, but reproducing the propulsive systems used by animals that can improve the versatility and response at low Reynolds number is indeed quite complex. The main objective of the present work is the computational study of the influence of the Reynolds number, frequency, and amplitude of the oscillatory movement of a NACA0012 airfoil in the aerodynamic performance. The thrust and power coefficients are obtained which together are used to calculate the propulsive efficiency. The simulations were performed using ANSYS Fluent with a RANS approach for Reynolds numbers between 8500 and 34,000, reduced frequencies between 1 and 5, and Strouhal numbers from 0.1 to 0.4. The aerodynamic parameters were thoroughly explored as well as their interaction, concluding that when the Reynolds number is increased, the optimal propulsive efficiency occurs for higher nondimensional amplitudes and lower reduced frequencies, agreeing in some ways with the phenomena observed in the animal kingdom.

Keywords: energy saving and efficiency; aerodynamic coefficients; propulsive efficiency; bioenergetics; biomimetics

---

1. Introduction

When the flapping airfoil/wing mechanism was unveiled as a thrust production system, the scientific community foresaw the possibility to investigate new aerodynamic phenomena and develop newer systems that could improve substantially the way airplanes fly nowadays, which is today still rather conservative [1]. However, in the first part of the 20th Century, very little effort was made in terms of understanding and exploiting the aerodynamics of living beings.

Animals such as insects [2], birds, small fishes, and even the big blue whale are equipped with a spectacular propulsion system that was subjected to natural selection processes over millions of years, which inevitably offers a significant advantage [3,4].

Based on the phenomena seen in Nature, Micro Aerial Vehicles (MAVs) and Nano Aerial Vehicles (NAVs) with indispensable civil and military applications such as surveillance, espionage, atmospheric weather monitoring, and catastrophe relief purposes [5] are being developed to offer undeniable maneuverability and efficiency at lower length scales. The most advanced MAVs related research projects define these vehicles as vehicles with no length, width, or height larger than 15 cm, as declared in a Defense Advanced Research Projects Agency (DARPA) program [6].

The flapping airfoil was firstly studied by Knoller and Verein [7] and Betz [8] that found that while plunging an airfoil, an effective angle of attack, which changes sinusoidally over time, would be created. As a result, an oscillatory aerodynamic force normal to the relative velocity was generated which could be decomposed in lift and thrust forces. Katzmayr [9] experimentally verified the Knoller–Betz effect in an interesting way, by placing a stationary airfoil into a sinusoidal wind stream and measuring an average thrust.

However, the Knoller and Betz theory was only based on the airfoil motion and did not account for the vorticity shed downstream of the airfoil. Thus, later in 1935, Kármán and Burgers [10] successfully explained theoretically the thrust generation mechanism based on the vortex shedding on the downstream side of the airfoil and the orientation of the wake vortices, identifying the typical von Kármán vortex street which is always associated with the production of drag or an averaged jet-like flow that generates a propulsive force explained by Newton's third law of motion.

All these concepts were further studied by Freymuth [11] using flow visualization of both plunging and pitching motions. Firstly, studying the aerodynamics of plunging and pitching airfoils offered a better understanding regarding the flutter and gust-response effects, which are based only on the analysis of the lifting forces [12,13]. Oscillating airfoils also opened new ways to study the impact of the dynamic stall on helicopter and propeller blades performance and how impactful is the wake created by a foregoing blade on the following blades [14].

Plunging airfoils have been also analyzed by Lai and Platzer [15], Lewin and Haj-Hariri [16], and Young [17], where the wake structures were intensively studied, such as the vortex-pair shedding that represents the transition from the drag producing wake to the thrust producing inverted von Kármán vortex street. Young and Lai [18] concluded that this type of wake structure was caused by the interaction between bluff-body type natural shedding from the trailing edge and the motion of the airfoil and, more recently, Andersen et al. [19] suggested that a drag-thrust transition wake showing a two-vortex pair configuration per oscillation period is a characteristic of low frequencies and high amplitudes oscillations.

The plunging motion was further investigated by Young and Lai [20,21] who found the impact of the several parameters governing this problem, especially the Strouhal number, which showed that maximum thrust and optimum efficiency take place at the near dynamic stall boundary. However, for lower Reynolds numbers, the problem becomes more complicated because efficient lift and thrust generation is achieved by shedding the leading-edge vortex [22].

Interested in finding which flying conditions animals operate in, Taylor et al. [23] dedicated their studies to forty-two species of birds, bats, and insects in cruise flight and verified that these creatures fly within a limited range of Strouhal numbers between 0.2 and 0.4. Hence, this parameter is a possible indicator of the flapping conditions that provide the most efficient flight, being essential to characterize the flight of several natural flyers, regardless of their scale.

Natural flyers use these forces effectively where thrust and lift production means that energy is transferred from the wings to the fluid. However, the study of oscillating airfoils has undergone a drastic shift when researchers identified the potential of flapping airfoils to extract energy from the flow field [24]. Studies on energy harvesting efficiency over oscillating airfoils revealed that a harvester could achieve maximum efficiency of about 30% on sinusoidal motion for $Re = 1000$ and 40% to 50% on nonsinusoidal motion; other efficiency-enhancing mechanisms might include corrugations at the foil surface, its structural flexibility, and multiple foil configuration [25] such as the parallel foil configuration numerically investigated in [26].

Introducing structural flexibility, as studied by Zhu et al. [27], the airfoil shows a superior capability regarding energy extraction when comparing to the rigid case. The same conclusion is obtained by Jeanmonod and Olivier [28], who firmly stated that a foil with constant mechanical properties in the chordwise direction is far from being an optimal solution, demonstrating that a flexible plate can extract double the power of a rigid structure.

Another way to improve extraction efficiency can also be achieved by creating an effective camber which will produce a lift enhancement that is associated with the output power. This concept was studied by Bouzaher et al. [29], who added a Gurney flap at the trailing-edge that when synchronized with the flapping motion at a $Re = 1100$ created this virtual camber that improved the output power. The lift production can be further improved by controlling the development of the leading-edge vortex than, when attached to the airfoil surface, contributed to the improvement in lift generation which brings once again an increase in the energy harvesting magnitude [24]. The same concept may be applied for thrust enhancement where camber morphing also reveals the potential to maximize the ratio of thrust to lift of flapping airfoils in several flight conditions [30].

Recently, Xia et al. [31] presented a wide range review on the fluid dynamics of flapping foils/wings where the authors reviewed the effects of some key parameters such as the Reynolds number, reduced frequency, and flapping amplitude. In addition, the intricacies of this problem are explored and the main challenges are identified, such as, highly efficient thrust mechanism, 3D flow control methods, and types of motion.

Regarding the type of motion, Boudis et al. [32] conclude that the sinusoidal waveform appears to favor the maximum propulsive efficiency notwithstanding the fact that nonsinusoidal trajectories can offer undeniable improvements, offering an increase up to 110% in terms of thrust production. At a Reynolds number of 5000, Dash et al. [33] also studied the influence of the airfoil motion concluding that, when at high frequencies, the thrust performance can be recovered when tweaking the effective angle of attack profile to a different type of wave, such as the square waveform.

The present work studied numerically the flapping airfoil problem, in particular, the plunging motion which has not been yet subjected to sufficiently detailed studies when compared with the combined plunging and pitching. Hence, the present work aims at studying the influence of flight velocity, the motion's frequency, and amplitude to fulfill this gap and help to understand the generation of thrust and what combinations in the operating domain are energetically adequate to a vehicle that uses the flapping mechanism as its mean of motion.

2. Methodology

The numerical methodology adopted for this paper is presented in this section. Firstly, the Reynolds-averaged Navier–Stokes (RANS) equations are selected as the governing equations, due to the unsteadiness of the current problem and the inherent turbulent phenomena seen in nature. The continuity and momentum averaged equations are written in a Cartesian tensor form as (e.g., ANSYS [34] or Lopes et al. [35]):

$$\frac{\partial \rho}{\partial t} + \frac{\partial (\rho u_i)}{\partial x_i} = 0 \tag{1}$$

and

$$\frac{\partial (\rho u_i)}{\partial t} + \frac{\partial (\rho u_i u_j)}{\partial x_j} = -\frac{\partial p}{\partial x_i} + \frac{\partial}{\partial x_j}\left[\mu \left(\frac{\partial u_i}{\partial x_j} + \frac{\partial u_j}{\partial x_i} - \frac{2}{3}\delta_{ij}\frac{\partial u_l}{\partial x_l} \right) \right] + \frac{\partial (-\rho \overline{u_i' u_j'})}{\partial x_j} \tag{2}$$

respectively.

The usage of an averaged formulation of the Navier–Stokes equations calls for the necessity of selecting a turbulence model that will model the velocity fluctuations as a function of the mean velocity field typically recurring to the Boussinesq hypothesis.

In the turbulence modeling field, some authors have studied the effectiveness of some turbulent models in representing the flow characteristics, but at the same time, some of the published research from other authors assumes the flow around flapping airfoils to be laminar. The Shear-Stress Transport $k - \omega$ model was selected to access its capability in predicting the flow surrounding a plunging airfoil since it shows a superior capability in representing the flow surrounding oscillatory airfoils [35].

In this study, the symmetrical airfoil NACA0012 is being used to simulate the plunging motion and analyze the flow configurations created by it. The airfoil's motion is described by the equation

$$y(t) = A\cos(2\pi f t) \tag{3}$$

and its velocity is given by

$$\dot{y}(t) = -2\pi f A \sin(2\pi f t) \tag{4}$$

where A and f are respectively the motion's amplitude and frequency, respectively.

Simulations were carried out using ANSYS Fluent with a mesh (Figure 1a) containing two main zones, which allowed the creation of a structured mesh around the airfoil (Figure 1b) and an unstructured grid in the airfoil's far-field. This mesh design effectively reduced the computational demand and time since only the outside part of the mesh was subjected to mesh update calculations such as deformation and remeshing.

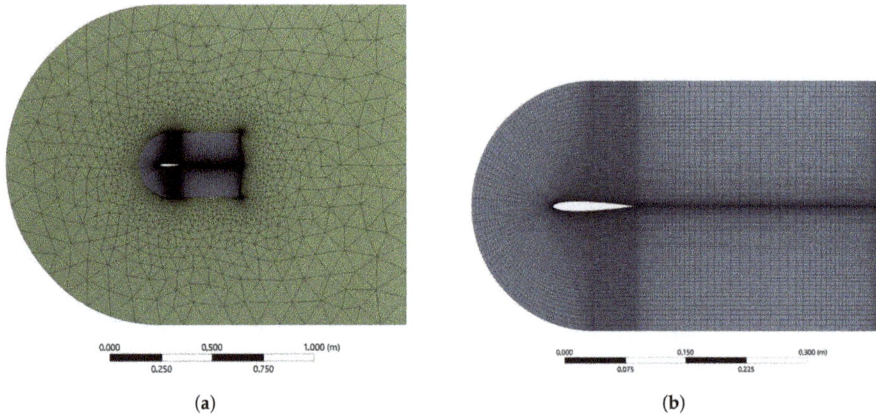

(**a**)   (**b**)

Figure 1. Computational mesh (**a**) and its internal zone (**b**).

The computational domain boundaries consisting of an inlet, an outlet, upper and lower walls, and the airfoil, is represented in Figure 2. The inlet was subjected to an inlet velocity boundary condition where the flow velocity was prescribed. On the outlet, the outflow boundary condition was applied, which is an extrapolation of the calculated variables on the interior. The remaining boundaries are walls, although with different characteristics, since the airfoil is treated as a wall where the no-slip condition is imposed, and the upper and lower walls are considered to have no shear stress which removes the boundary layer effects.

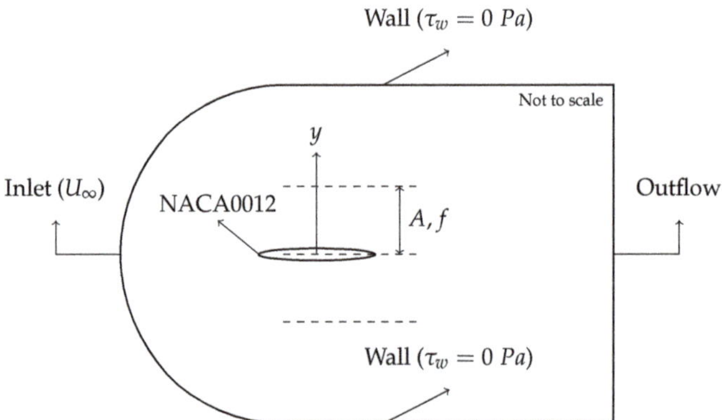

Figure 2. Computational domain.

The Least Squares Cell-Based scheme was selected as the gradients evaluation method, PREssure STaggering Option (PRESTO!) was chosen as the pressure interpolation arrangement, and the momentum, turbulent kinetic energy, and specific dissipation rate equations were discretized in a 2D space using the second-order accurate QUICK scheme. Regarding the transient formulation and knowing beforehand that the dynamic mesh feature was activated, the first order implicit method (in time) was the only scheme available. The pressure-velocity coupling algorithm used is the

Pressure-Implicit with Splitting of Operators (PISO) that is highly recommended for all transient flow calculations.

After each initialization, the mesh was adapted to ensure that y^+ was between 0 and 1 (values up to 5 were acceptable), since when using the $k - \omega$ turbulence model, it is fundamental to guarantee that the effects of the flow close to the walls are well resolved. Due to the boundary conditions, solution initialization and nonlinearity, which characterizes the Navier–Stokes equations, it was observed that the flow configuration would stabilize after analyzing at least five periods, always maintaining all residuals below 10^{-3}.

The processing and analysis of the data (drag and lift coefficients exported during simulations) were made recurring to in-house C-programs, whose aim was to obtain the averaged aerodynamic coefficients and propulsive efficiency.

As mentioned before, the flapping mechanism is typically correlated with thrust production since it can, in some conditions, positively change the wake momentum and because of that, it is important to evaluate not only the generated force but also the propulsive efficiency which is the ratio between the generated power, $-DU_\infty$, and the power supply is given by $-L\dot{y}$. The propulsive efficiency is then defined as

$$\eta = \frac{\overline{C_t}U_\infty}{\overline{C_P}} \tag{5}$$

where

$$\overline{C_t} = -\frac{1}{\Delta t} \int_t^{t+\Delta t} C_d \, dt \tag{6}$$

and

$$\overline{C_P} = -\frac{1}{\Delta t} \int_t^{t+\Delta t} \dot{y} C_l \, dt \tag{7}$$

In these two last equations, $\overline{C_t}$ is the mean thrust coefficient, C_d is the drag coefficient, C_l is the lift coefficient, and $\overline{C_P}$ is the mean power coefficient, which are parameters vastly used by researchers to evaluate the aerodynamic performance [24].

The influence of the mentioned parameters will be evaluated using the typical dimensionless quantities for a flapping airfoil: the reduced frequency, k, the nondimensional amplitude, h, Strouhal number, St, and the Reynolds number, Re, shown in Table 1 (see [31] for details). The variables U_∞, ρ, μ, c, f, and A are the inlet speed, air density, air dynamic viscosity, aerodynamic chord, motion frequency, and amplitude, respectively, all in SI units.

Table 1. Dimensionless parameters.

Reynolds Number, Re	Reduced Frequency, k	Nondimensional Amplitude, h	Strouhal Number, St
$Re = \dfrac{\rho U_\infty c}{\mu}$	$k = \dfrac{2\pi f c}{U_\infty}$	$h = \dfrac{A}{c}$	$St = \dfrac{2fA}{U_\infty}$

The numerical validation started by performing a boundary location, mesh, and time step independence studies for a Reynolds number of 17,000, a nondimensional amplitude of 0.5, and a reduced frequency of 2.5. The boundary location study, presented in Figure 3, focused on the analysis of the blockage ratio influence on the drag coefficient over time. In this paper, the blockage ratio, BR, was defined as being the ratio between the wake dimension and the inlet height, and it is seen that no considerable influence is seen for the different cases tested.

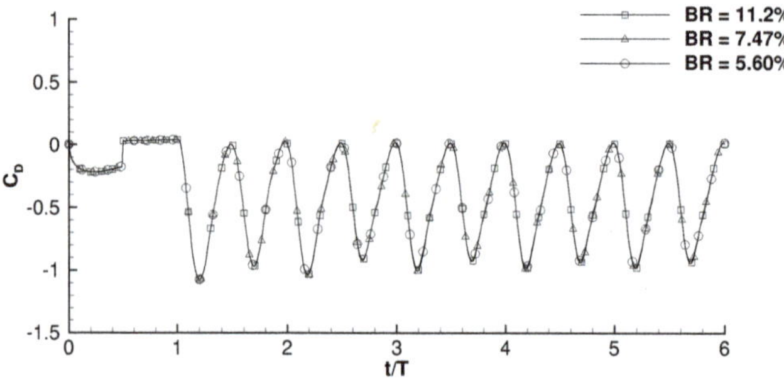

Figure 3. Boundary location study.

The mesh independence study was conducted through the refinement of the internal zone of the mesh into several divisions (50, 71, and 100 divisions), as shown in Figure 4. It can be concluded that the mesh with 71 internal divisions holds an independent result that corresponds to a global mesh with 62,469 cells/52,394 nodes.

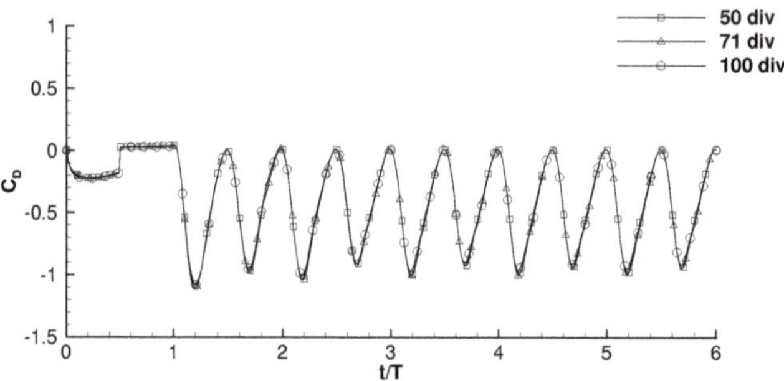

Figure 4. Mesh independence study.

The last phase focuses on the independence of the time step and for this study, three time steps were considered, calculated as $T/100$, $T/200$, and $T/400$ where T is the motion period. Figure 5 shows that a valid result is obtained for 200 points per oscillation period.

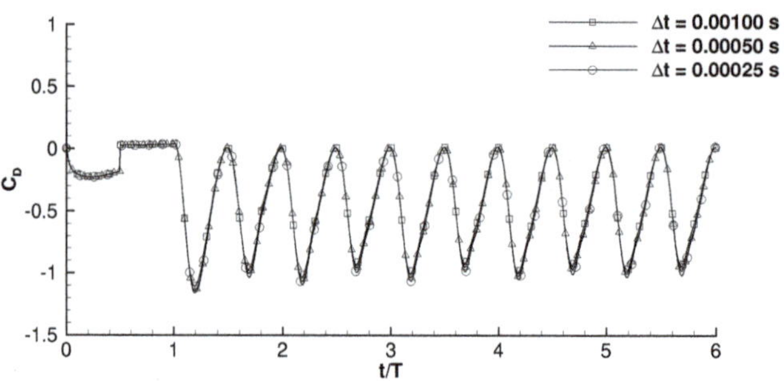

Figure 5. Time step independence study.

3. Results

This section shows the results concerning the influence of Reynolds number, motion's amplitude, frequency, and the Strouhal number on the flapping mechanism. Graphics of thrust and power coefficients, as well as propulsive efficiency, will be shown as functions of these relevant parameters. The pure plunging NACA0012 airfoil was firstly tested with a $Re = 8500$ with $1 \leq k \leq 5, 0.1 \leq St \leq 0.4$, and h never surpassing 0.5.

In Figure 6, the thrust coefficient is shown in the kh plane. In the contour plot, it is noteworthy that the C_t value is growing faster than the Strouhal number with respect to k, a phenomenon that becomes evident for a nondimensional amplitude higher than 0.3. The same behavior is not exhibited by the power coefficient, which has its isolines almost parallel to the ones that represent a constant Strouhal number.

It is important to note that there is an intimate relationship between the Strouhal number and the maximum effective angle of attack to which the airfoil is subject, so it is expected that the required power (C_P) and the maximum angle of attack also have one.

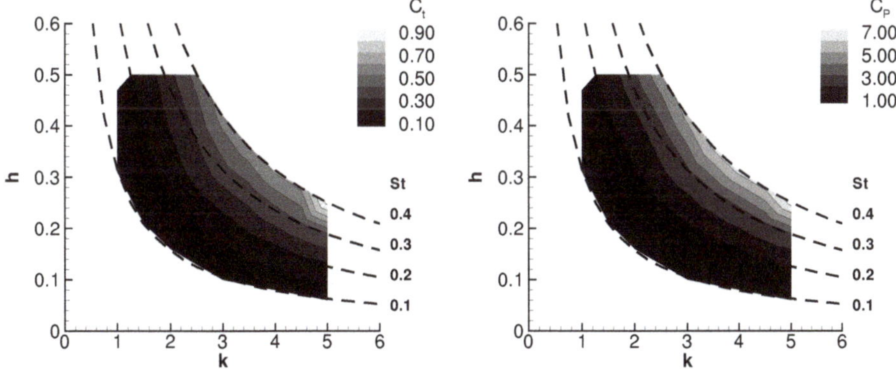

Figure 6. Mean thrust and power coefficients with $Re = 8500$.

Regarding the influence of the Strouhal number on the thrust and power coefficients, the following correlations are obtained:

$$C_t (St) = -0.051 + 0.462St + 2.668St^2 \ (R^2 = 0.87) \tag{8}$$

$$C_P (St) = 0.543 - 8.471St + 52.58St^2 \ (R^2 = 0.96) \tag{9}$$

with $0.1 \leq St \leq 0.4$. The coefficient of correlation for the C_t approximation is under 0.90 which once again reinforces the fact that the isolines of the thrust coefficient are not entirely parallel.

The consolidation of C_t and C_P results in the propulsive efficiency shown in Figure 7, which has a peculiar distribution in the kh plane since it does not depend only on the reduced frequency or nondimensional amplitude alone but rather on the combination of both. Thus, the maximum efficiency region is encountered in the vicinity of a Strouhal number of 0.15 that unfortunately is incompatible with the maximum thrust coefficient area, reaching a maximum value of 0.23.

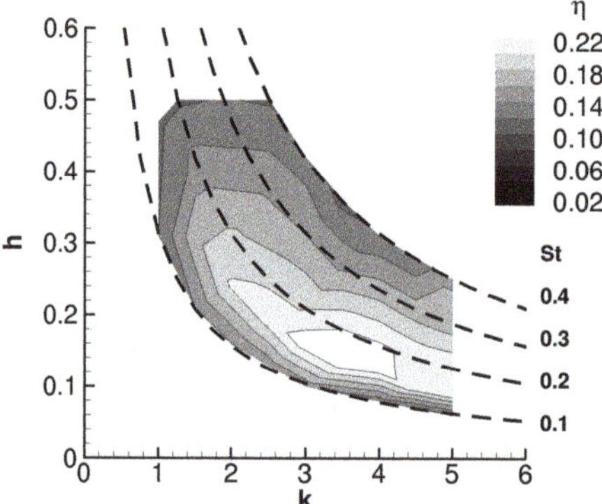

Figure 7. Propulsive efficiency with $Re = 8500$.

This, and the following propulsive efficiency graphics, are important to understanding how the energy given to the system is converted into the production of thrust (propulsive power) by an airfoil performing the plunging movement. Although it is already known that this sinusoidal movement may not represent the most efficient motion, it is nevertheless interesting to understand the regions where there is an optimal conversion of the input given to the system.

The airfoil was also tested with a Reynolds number of 17,000, keeping the previously mentioned parameters in the same range. The results show a similar distribution of the thrust coefficient, power coefficient (Figure 8), and propulsive efficiency, shown in Figure 9.

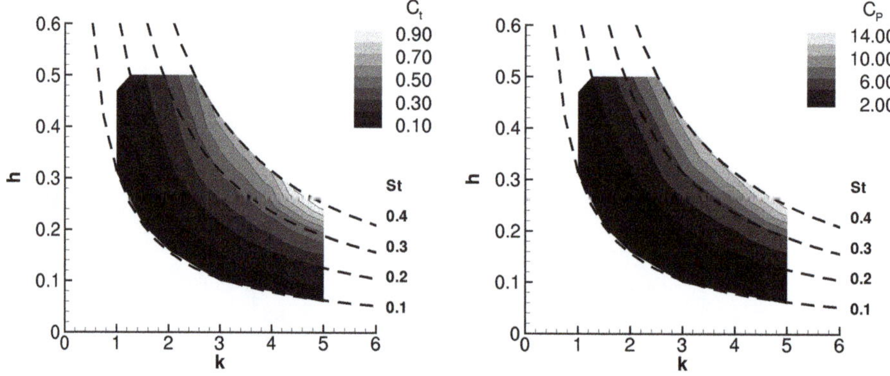

Figure 8. Mean thrust and power coefficients with $Re = 17,000$.

In respect to the influence of the Strouhal number on the thrust and power coefficients for a Reynolds number of 17,000, the following correlations were obtained:

$$C_t(St) = -0.029 + 0.312St + 3.584St^2 \ (R^2 = 0.88) \tag{10}$$

$$C_P(St) = 1.371 - 20.14St + 113.0St^2 \ (R^2 = 0.96) \tag{11}$$

with $0.1 \leq St \leq 0.4$.

The propulsive efficiency increase is the major difference, which may indicate that the plunging movement tends to be more efficient at higher Reynolds numbers.

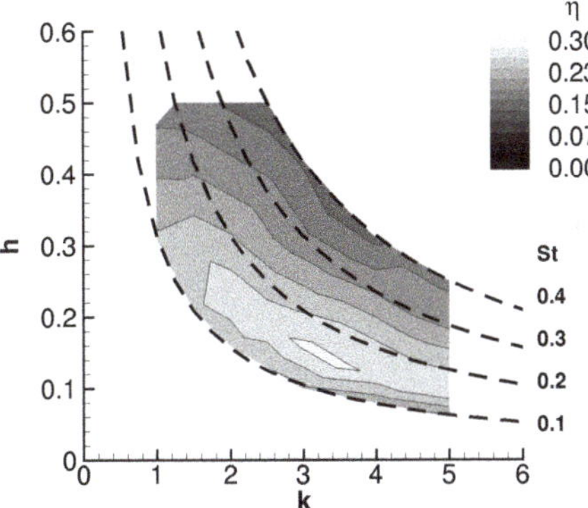

Figure 9. Propulsive efficiency with $Re = 17{,}000$.

In order to better interpret the evolution of propulsive efficiency for this specific case, propulsion efficiency curves are presented in Figure 10, considering the dimensionless amplitude (Figure 10a) and reduced frequency (Figure 10b) constant. When looking at the graphics, it is possible to verify that the maximum propulsive efficiency reached occurs at higher reduced frequencies when the dimensionless amplitude decreases. Regarding the graphic on the right, similar behavior is verified, since the highest propulsive efficiencies are detected at the highest reduced frequencies and lowest dimensionless amplitudes. However, a deeper comparison between both graphs concludes that the conversion of the required energy associated with the movement into propulsive energy is highly sensitive to the change in the nondimensional amplitude when a constant reduced frequency is considered, as observed by the higher slopes verified in the graphic on the right.

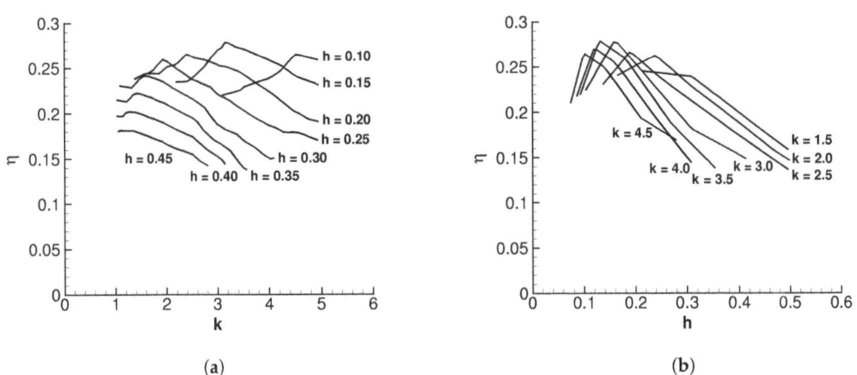

(a)

(b)

Figure 10. Propulsive efficiency with $Re = 17{,}000$ assuming constant h (**a**) and k (**b**).

The Reynolds number was further increased to a maximum tested value of 34,000 (Figure 11). In this operating regime, the upper limit of the Strouhal range was limited to 0.2 since no clear advantage was seen in terms of achieving better aerodynamic performance. At this operating condition, an interesting fact is that the thrust coefficient isolines become equidistant to the constant St curves. In this situation, it becomes evident that the isolines of C_t, St, and C_P are all doubtlessly parallel. Due to this turn of events, the maximum propulsive efficiency region was translated to the left, as seen in Figure 12 and, because of that, an additional zone ($k < 1$) was considered to understand the aerodynamic performance in low reduced frequencies.

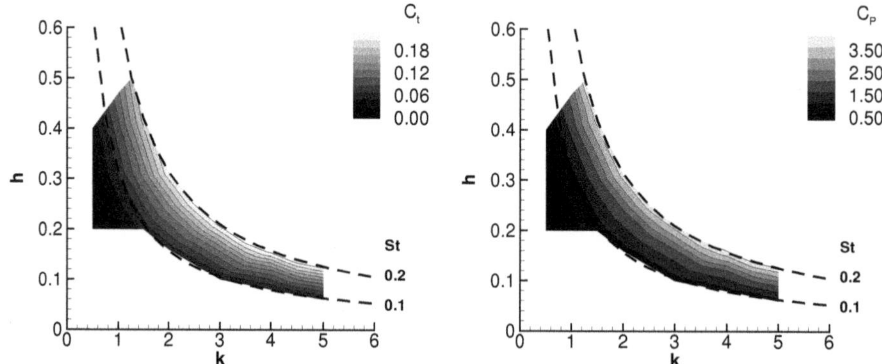

Figure 11. Mean thrust and power coefficients with $Re = 34,000$.

The influence of the Strouhal number on the thrust and power coefficients for the actual number of Reynolds is also studied, obtaining the correlations:

$$C_t\,(St) = -0.059 + 0.671St + 3.643St^2 \;(R^2 = 0.96) \tag{12}$$

$$C_P\,(St) = 0.587 - 12.71St + 146.5St^2 \;(R^2 = 0.99) \tag{13}$$

with $0.1 \leq St \leq 0.2$. The coefficient of determination regarding the C_t approximation improved $(R^2 > 0.95)$ in comparison with the previously tested numbers of Reynolds, which corroborates the fact that the isolines of C_t became parallel with the Strouhal number.

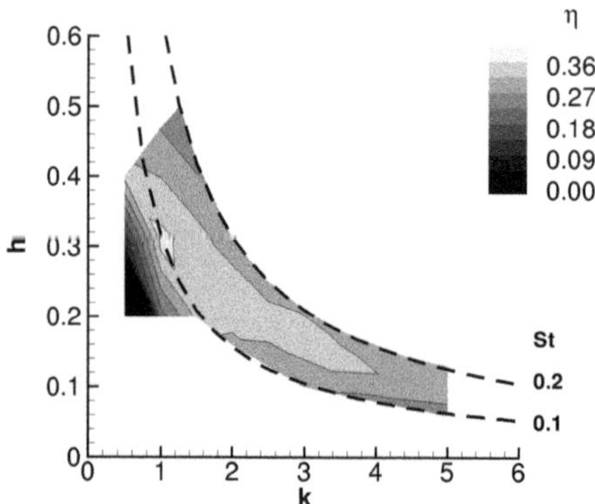

Figure 12. Propulsive efficiency with $Re = 34,000$.

Overall, from a Reynolds number of 8500 to 34,000, the propulsive efficiency increased in comparison to the previous cases, which again suggests that the pure plunging motion is favored while operating at relatively high Reynolds numbers. Although animals use plunging and pitching combined, the results presented in this section are very much in concordance with what is seen in nature, considering that smaller animals tend to operate at high reduced frequencies and low nondimensional amplitudes. In comparison, bigger animals do the exact opposite.

4. Conclusions

Nature has been the main source of concepts that inspire the systems developed by engineers who, since the beginning of time, understood that it is powered by evolution mechanisms that tend to offer optimized solutions regarding the environmental conditions. These mechanisms directly affect animals, from the smallest insect to the big blue whale, that over millions of years, made them very well adapted to their habitat and to the way they move.

In this work, the flapping airfoil problem was investigated, utilizing a NACA0012, by studying the influence of several parameters such as motion frequency and amplitude, the Reynolds number, and the Strouhal number. Thrust, lift, and power coefficients, as well as the propulsive efficiency, were the selected parameters to analyze the aerodynamic performance.

The results indicate that the power coefficient isolines demonstrated to be almost parallel to the hyperbolas representing constant Strouhal numbers, a phenomenon not verified for the thrust coefficient, except for the $Re = 34{,}000$ case. It was also seen that higher frequencies and amplitudes propitiate higher thrust forces as well as required power while, in terms of propulsive efficiency, for the $Re = 8500$ and $17{,}000$ cases, higher reduced frequencies and lower amplitudes are preferred and for the $Re = 34{,}000$ case, higher amplitudes and lower reduced frequencies are favored. The results highlight that although a constant St gives an infinity of combinations (k, h), it is seen as an interesting parameter that offers some correlations about the aerodynamic performance, especially at higher Reynolds numbers.

The physical phenomena underlying the flapping airfoil and flapping wing mechanisms were extensively studied in recent years, which resulted in a greater insight regarding the thrust generation. Nevertheless, much more investigation is needed to efficiently develop and produce vehicles with these propulsive systems for any stage of flight. Ways to improve knowledge should be focused on testing different geometries in order to understand how geometrical parameters such as the aerodynamic chord, camber, and thickness can influence the aerodynamic performance. Another way to boost the bio-inspired design is on the materials side, seeking innovative and different properties (flexibility, porosity) that may improve the flapping mechanism and bring us closer to developing animal-like propulsive/lifting systems and structures.

Author Contributions: Conceptualization, E.C. and J.B.; Data curation, E.C.; Formal analysis, E.C.; Methodology, E.C. and A.S.; Project administration, A.S. and J.B.; Resources, A.S.; Software, E.C.; Supervision, A.S. and J.B.; Validation, E.C. and A.S.; Visualization, E.C.; Writing—original draft, E.C. and F.N.; Writing—review & editing, A.S. and J.B. All authors have read and agreed to the published version of the manuscript.

Funding: The present work was performed under the scope of the Aeronautics and Astronautics Research Center (AEROG) of the Laboratório Associado em Energia, Transportes e Aeronáutica (LAETA) activities and it was supported by Fundação para a Ciência e Tecnologia (FCT) through project numbers UID/EMS/50022/2019 and UIDB/50022/2020.

Conflicts of Interest: The authors declare no conflict of interest. The funders had no role in the design of the study; in the collection, analyses, or interpretation of data; in the writing of the manuscript, or in the decision to publish the results.

References

1. Lee, J.-S.; Kim, C.; Kim, K.H. Design of Flapping Airfoil for Optimal Aerodynamic Performance in Low-Reynolds. *AIAA J.* **2006**, *44*, 1960–1972. [CrossRef]
2. Barata, J.M.M.; Neves, F.M.S.P.; Manquinho, P.A.R. Comparative Study of Wing's Motion Patterns on Various Types of Insects on Resemblant Flight Stages. In Proceedings of the AIAA Science and Technology Forum 2015, Kissimmee, FL, USA, 5–9 January 2015.
3. Barata, J.M.M.; Manquinho, P.A.R.; Neves, F.M.S.P.; Silva, T.J.A. Propulsion for Biological Inspired Micro-Air Vehicles (MAVs). In Proceedings of the ICEUBI, International Conference on Engineering, Engineering for Society, Covilhã, Portugal, 2–4 December 2015.
4. Barata, J.M.M.; Manquinho, P.A.R.; Neves, F.M.S.P.; Silva, T.J.A. Propulsion for Biological Inspired Micro-Air Vehicles (MAVs). *Open J. Appl. Sci.* **2016**, *66*, 7–15. [CrossRef]
5. Vuruskan, A.; Fenercioglu, I.; Cetiner, O. A study on forces acting on a flapping wing. *EPJ Web Conf.* **2013**, *45*, 01028. [CrossRef]

6. Platzer, M.F.; Jones, K.D.; Young, J.; Lai, J.C.S. Flapping Wing Aerodynamics—Progress and Challenges. *AIAA J.* **2008**, *46*, 2136–2149. [CrossRef]

7. Knoller, R.; Verein, O.F. *Die Gesetze des Luftwiderstandes*; Verlag des Osterreichischer Flugtechnischen Vereines: Wien, Austria, 1909.

8. Betz, A. Ein Beitrag zur Erklaerung des Segelfluges. *Zeitschrift fur Flugtechnik und Motorluftschiffahrt* **1912**, *3*, 269–272.

9. Katzmayr, R. Effect of Periodic Changes of Angle of Attack on Behavior of Airfoils, NACA Report 147. 1922. Available online: https://digital.library.unt.edu/ark:/67531/metadc55795/m1/1/ (accessed on 10 April 2020).

10. von Kármán, T.; Burgers, J.M. General Aerodynamic Theory—Perfect Fluids. In *Aerodynamic Theory*; Durand, W.F., Ed.; Springer: Berlin, Germany, 1934; Volume 26.

11. Freymuth, P. Propulsive Vortical Signatures of Plunging and Pitching Airfoils. *AIAA J.* **1988**, *26*, 881–883. [CrossRef]

12. Jones, K.D.; Dohring, C.M.; Platzer, M.F. Experimental and Computational Investigation of the Knoller-Betz Effect. *AIAA J.* **1998**, *36*, 1240–1246. [CrossRef]

13. Koochesfahani, M.M. Vortical Patterns in the Wake of an Oscillating Airfoil. *AIAA J.* **1989**, *27*, 1200–1205. [CrossRef]

14. Tuncer, I.H.; Platzer, M.F. Computational Study of Flapping Airfoil Aerodynamics. *AIAA J. Aircr.* **2000**, *37*, 514–520. [CrossRef]

15. Lai, J.C.S.; Platzer, M.F. Jet Characteristics of a Plunging Airfoil. *AIAA J.* **1999**, *37*, 1529–1537. [CrossRef]

16. Lewin, G.C.; Haj-Hariri, H. Modelling Thrust Generation of a Two-Dimensional Heaving Airfoil in a Viscous Flow. *J. Fluid Mech.* **2003**, *492*, 339–362. [CrossRef]

17. Young, J. Numerical Simulation of the Unsteady Aerodynamics of Flapping Airfoils. Ph.D. Thesis, The University of New South Wales, Sydney, Australia, 2005.

18. Young, J.; Lai, J.C.S. Vortex Lock-in Phenomenon in the Wake of a Plunging Airfoil. *AIAA J.* **2007**, *45*, 485–490. [CrossRef]

19. Andersen, A.; Bohr, T.; Schnipper, T.; Walther, J. Wake structure and thrust generation of a flapping foil in two-dimensional flow. *J. Fluid Mech.* **2017**, *812*, R4. [CrossRef]

20. Young, J.; Lai, J.C.S. Oscillation Frequency and Amplitude Effects on the Wake of a Plunging Airfoil. *AIAA J.* **2004**, *42*, 2042–2052. [CrossRef]

21. Young, J.; Lai, J.C.S. Mechanisms Influencing the Efficiency of Oscillating Airfoil Propulsion. *AIAA J.* **2007**, *45*, 1695–1702. [CrossRef]

22. Gursul, I.; Cleaver D. Plunging Oscillations of Airfoils and Wings: Progress, Opportunities and Challenges. *AIAA J.* **2019**, *57*, 3648–3665. [CrossRef]

23. Taylor, G.K.; Nudds, R.L.; Thomas, A.L.R. Flying and Swimming Animals Cruise at a Strouhal Number Tuned for High Power Efficiency. *Nature* **2003**, *425*, 707–711. [CrossRef]

24. Geissler, W. Flapping Wing Energy Harvesting: Aerodynamic Aspects. *CEAS Aeronaut. J.* **2019**. [CrossRef]

25. Xiao, Q.; Zhu, Q. A review on flow energy harvesters based on flapping foils. *J. Fluids Struct.* **2014**, *46*, 174–191. [CrossRef]

26. Wang, Y.L.; Jiang, W.; Xie, Y. H. Numerical Investigation Into the Effects of Motion Parameters on Energy Extraction of the Parallel Foils. *ASME. J. Fluids Eng.* **2019**, *141*, 061104. [CrossRef]

27. Zhu, B.; Zhang, W.; Huang, Y. Energy extraction properties of a flapping wing with a deformable airfoil. *IET Renew. Power Gener.* **2019**, *13*, 1823–1832. [CrossRef]

28. Jeanmonod, G.; Olivier, M. Effects of chordwise flexibility on 2D flapping foils used as an energy extraction device. *J. Fluids Struct.* **2017**, *70*, 327–345. [CrossRef]

29. Bouzaher, M.T.; Drias, N.; Guerira, B. Improvement of Energy Extraction Efficiency for Flapping Airfoils by Using Oscillating Gurney Flaps. *Arab. J. Sci. Eng.* **2019**, *44*, 809–819. [CrossRef]

30. Soni, P.; Bharadwaj, A.; Ghosh, S. Optimization of the Thrust to Lift Ratio using Camber Morphing in Flapping Airfoil. In Proceedings of the AIAA Scitech 2020 Forum, Orlando, FL, USA, 6–10 January 2020.

31. Wu, X.; Zhang, X.; Tian, X.; Li, X.; Lu, W. A review on fluid dynamics of flapping foils. *Ocean. Eng.* **2020**, *195*, 106712. [CrossRef]

32. Boudis, A.; Bayeul-Lainé, A.C.; Benzaoui, A.; Oualli, H.; Guerri, O.; Coutier-Delgosha, O. Numerical Investigation of the Effects of Nonsinusoidal Motion Trajectory on the Propulsion Mechanisms of a Flapping Airfoil. *ASME J. Fluids Eng.* **2019**, *141*, 041106. [CrossRef]
33. Dash, S.M.; Lua, K.B.; Lim, T.T. Thrust Enhancement on a Two Dimensional Elliptic Airfoil in a Forward Flight. *Int. J. Mech. Aerosp.* **2016**, *10*, 265–272.
34. ANSYS Inc. *Reynolds (Ensemble) Averaging, Fluent Theory Guide, 2019, Fluent 2019 R3*; ANSYS Inc.: Canonsburg, PA, USA, 2019.
35. Lopes, R.; Camacho, E.; Neves, F.; Silva, A.; Barata, J. Numerical and Experimental Study of a Plunging Airfoil. In Proceedings of the 4th Thermal and Fluids Engineering Conference, Las Vegas, NV, USA, 14–17 April 2019.

Article

Aircraft Propellers—Is There a Future? [†]

Pedro Alves *, Miguel Silvestre and Pedro Gamboa

C-MAST—Center for Mechanical and Aerospace Science and Technologies, University of Beira Interior, Rua Marquês d'Ávila e Bolama, 6201-001 Covilhã, Portugal; mars@ubi.pt (M.S.); pgamboa@ubi.pt (P.G.)
* Correspondence: ferreira.alves@ubi.pt
†   This paper is an extended version of our paper published in ICEUBI2019—International Congress on Engineering—Engineering for Evolution, Covilhã, Portugal, 27–29 November 2019.

Received: 18 June 2020; Accepted: 7 August 2020; Published: 11 August 2020

Abstract: The race for speed ruled the early Jet Age on aviation. Aircraft manufacturers chased faster and faster planes in a fight for pride and capability. In the early 1970s, dreams were that the future would be supersonic, but fuel economy and unacceptable noise levels made that era never happen. After the 1973 oil crisis, the paradigm changed. The average cruise speed on newly developed aircraft started to decrease in exchange for improvements in many other performance parameters. At the same pace, the airliner's power-plants are evolving to look more like a ducted turboprop, and less like a pure jet engine as the pursuit for the higher bypass ratios continues. However, since the birth of jet aircraft, the propeller-driven plane has lost its dominant place, associated with the idea that going back to propeller-driven airplanes, and what it represents in terms of modernity and security, has started a propeller avoidance phenomenon with travelers and thus with airlines. Today, even with the modest research effort since the 1980s, advanced propellers are getting efficiencies closer to jet-powered engines at their contemporary typical cruise speeds. This paper gives a brief overview of the performance trends in aviation since the last century. Comparison examples between aircraft designed on different paradigms are presented. The use of propellers as a reborn propulsive device is discussed.

Keywords: propeller; aircraft; turboprop; flight efficiency; flight speed

---

1. Introduction

The propeller is a device that converts the rotary power of an engine or motor into a thrust force that pushes the vehicle to which it is attached. Comprised by one or more radial airfoil-section blades rotating about an axis, the propeller acts as a rotating wing. Aircraft propellers first emerged at the end of the 18th century; however, this study only discusses its history from the 20th century and beyond. See Ref. [1,2] for a historical review from the preceding decades. By the end of the 19th century, a feeling of disbelief on heavier-than-air manned flight was present [3]. The first controlled, powered flight, starred by the Wright Brothers in 1903, marked the turn of a page of skepticism concerning heavier-than-air manned flight. This remarkable achievement brought an increased excitement around the aviation community, and in the period 1905–1910, there was an impressive growth in the number of filed patents [4] (see Figure 1).

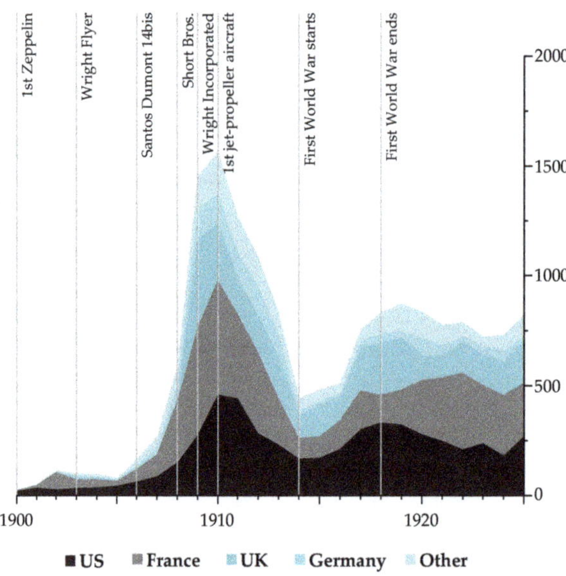

Figure 1. First patent filings by origin, 1900–1925. Between 1900 and 1970, patent filings relating to aviation tended to concentrate in the US, France, Germany, and the UK. Source: adapted from [4].

This pre-WWI (World War I) period was also responsible for a transition from individuals as hobbyists and enthusiasts, motivated by curiosity, pride, and fame, to institutions and governments acknowledging the airplanes as a strategic weapon to win wars. By the end of WWI, from the 1920s to the 1930s, designers, engineers, and inventors established new, active, and venturous aeronautic communities in Europe and North America. This prosperous era of innovation and technological growth of aviation extended its developments to all components of the airplane, including the propeller. Donald W. Douglas, head of the Douglas Aircraft Company, considered those communities of people responsible for helping change the world, acknowledging propeller makers and their creations indispensable for success [5]. The work on those propulsive devices joined the higher power outputs of the newer engines to the improved body aerodynamics resulting in higher performance aircraft capable of "climb quicker and cruise faster using less power and if need be, fly to safety on one engine" [3]. Since the first effective propellers powered by piston engines, through impressive supersonic aircraft and up to modern airliners, a lot has changed in aviation. The aircraft is now a balance between hundreds of different specifications. Some being improved at the cost of others. Today, the advent of electric multirotor vertical take-off and landing aircraft [6–11], starting as unmanned aerial vehicles but also aimed for personal transportation are bringing the assertion of the propeller as the main choice for low-speed, state-of-the-art efficient propulsion devices. Nevertheless, propellers are nothing more than a niche in commercial aviation commute aircraft. However, this century brought new challenges and priorities. Global warming, pollution, and sustainability are now serious concerns [12]. The present work shows the evolution of cruise speeds, especially on commercial aviation, in the past century to realize that the propeller comeback may be the next innovation towards more sustainable aviation. The trends are presented, and their motives are discussed. In the first section, a brief historical overview is presented. The second section introduces the influence of the cruise speed in the aircraft aerodynamic efficiency and engine fuel consumption. Then it discusses the evolution of the flight speed of the airliners since the jet age. The third section shows the relevance of bringing back the propeller and continue its development.

2. Early Jet Age: The Race for Speed

With the invention and development of the jet engine during WWII, gas turbine-powered aircraft expanded the whole flight envelope. Flying higher and faster, both commercial and military jet airplanes ruled the 1950s and 1960s, at what was called: The Jet Age [13]. Compared with piston airplanes, the speed and ceiling of these first jet-powered aircraft were incredible, and the race for flight speed became the leading trend [3]. The following years gave birth to a generation of even faster jet aircraft as the example of the Boeing 727, which had a maximum cruise speed of Mach 0.84. However, to overcome the speed of sound, a larger amount of power was required. This is due to the sharp rise in drag, experienced above a critical Mach number [14] (Figure 2). Also, supersonic flight introduced the need for pure jet engines since the propeller blades encounter the critical Mach number at smaller cruise speeds due to their additional rotation speed.

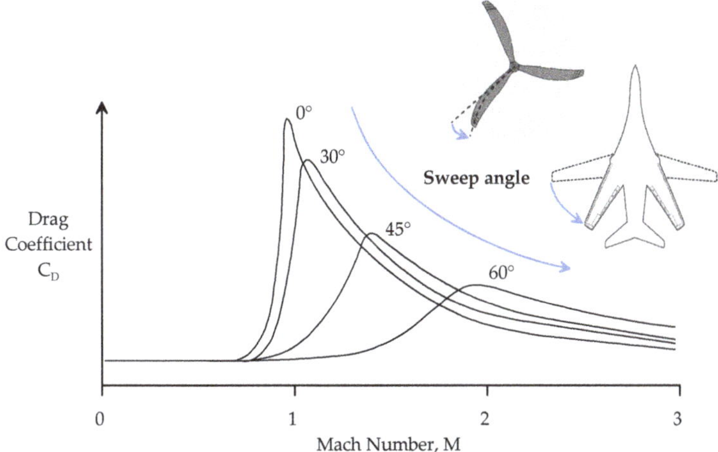

Figure 2. Drag rise due to cruise Mach and the effect of wing sweep. Adapted from [15].

Therefore, the race continued, and in the early 1960s, the Convair 990 could already fly at speeds up to Mach 0.87 [16]. At this time, in one of the test flights, Douglas company accelerated its DC-8 to Mach 1.01 in a 16-second dive. Nevertheless, commercial aviation did not stop there. Pursuing military achievements as the Lockheed SR-71 "Blackbird", that cruised at Mach 3.2, commercial aviation needed to rush forward. In the early 1970s, jet engine technology was developing at a tremendous pace. Commercial aircraft, characterized by sharp noses and high swept wings, also started breaking through the sound barrier. The Concorde and the Tupolev Tu-144 were developed to cruise at Mach 2. However, Mach 2 still seemed not enough, so, Boeing wanted to create an even faster commercial airplane [17–19], aiming to fly at Mach 3. Suddenly hypersonic transportation was the subject of research [20,21]. It seemed that there were no limits to the race for speed, but something went wrong, and the true supersonic age never came. The magnificence of supersonic airliners was comparable only to the horror of their ecology and economy. The supersonic engines roar was annoying to the cities' populations. Sound booms were destroying everything around [22], leading many countries to ban supersonic flights from their airspace [23]. In addition to that, the super-powerful afterburner engines were so hungry for fuel that airlines had to increase the cost of tickets to cover their expenses. Those facts, combined with the 1970s oil crisis, made the supersonic flight not to enter mass aviation. To transport ordinary travelers on such planes was the same as the average citizen commuting to work on supercars. One of the main reasons that made the airlines and manufacturers abandon the race for speed is also one of the main elements of the aircraft: the engine. The heart of most modern aircraft is a jet engine. The task of any jet engine (or reaction engine) is to convert the fuel's chemical energy into the jet flow's kinetic energy. In practice, the fuel ignites, expands, accelerates, and pushes the machine forward. The jet engines used in aviation use not only their fuel but also the surrounding air, which is also heated up and accelerated to be ejected at high speed by a nozzle to create thrust. See Ref. [24] for further insights on jet engines. The Rolls-Royce / Snecma Olympus 593 that equipped the Concorde is a classic example turbojet engine. However, these engines were very greedy. With its small capacity, Concorde had a huge fuel consumption, resulting from the great increase in drag shown in Figure 2. The much larger Boeing 747, produced in 1968, turned out to be much more economical [25]. Despite that, new larger subsonic jet-powered airliners conquered the main long-haul routes, and smaller models were conquering the regional ones.

3. Fuel Efficiency

For commercial aircraft, fuel efficiency is usually regarded as the inverse of the fuel consumption, which is fuel quantity burned per unit traveled distance per unit passenger, normally, in $kmPax/L_{Fuel}$. When comparing different fuels, fuel mass is more meaningful than fuel volume. If, instead of mass or volume of fuel, the amount of energy consumed (or contained in that mass of fuel) was used, one could even compare different types of propulsion systems, e.g., the electric propulsion. Considering the propulsive system efficiency as $\eta_{th}\eta_p$, where η_{th} is the thermal efficiency and η_p is the propulsive efficiency, and that in cruise, the required thrust is:

$$F = \frac{W}{L/D} \tag{1}$$

where F is the propulsive thrust force, W is the aircraft weight, L and D are aircraft lift and drag, respectively.

Using Equation (1), the aircraft fuel efficiency can be regarded as:

$$\eta_{Fuel} = \frac{L}{D}\frac{pax}{W}\eta_{th}\eta_p \tag{2}$$

Therefore, the aircraft will be fuel-efficient if it has large aerodynamic efficiency (L/D) and low take-off weight per unit passenger (W/pax). The aerodynamic efficiency depends mostly on the airfoil design and wing design parameters: span; chord; sweep; etc., and all other elements of the aircraft that generate drag but not produce any lift. The weight per unit passenger depends mostly on the aircraft structure, materials, and design. The engine thermal efficiency has been improving in recent decades being close to its upper limits. Regarding the propulsive efficiency, to propel the aircraft, the propulsive system generates the thrust by accelerating the incoming air stream mass flow from V_{Cruise} to a propulsive stream with V_{Jet}, considering that the fuel mass flow is much smaller than this propulsive stream, we get Equation (3).

$$\eta_p = \frac{2}{2 + \dfrac{F}{\dot{m}V_{Cruise}}} \tag{3}$$

with $\dot{m}$ being the propulsive stream mass flow rate.

According to Equation (3), the propulsive efficiency increases if $\dot{m}$ is increased. This was accomplished in the jet engine by adding an external bypass outer stream of cold air mass flow. These engines were named turbofans and became a real classic solution in modern aviation. Since their birth, turbofan engines started to replace turbojets, becoming as fast as them, despite that the advantage of adding the bypass cold stream diminishes towards higher V_{Cruise}. Most modern fighters like the F-15, Eurofighter Typhoon, Sukhoi Su-30, and the newest F-22 and Sukhoi Su-57 are equipped with turbofan engines. At V_{Cruise} of current commercial aircraft, the bypass stream fan is in fact a ducted propeller. Figure 3 shows the typical propulsive efficiencies for the most common aircraft engine types. The influence in the propulsive efficiency of parameters such as the propeller blade sweep and engine bypass ratio is also represented. It is clear that the bypass ratio of the turbofan increases the propulsive efficiency relative to the turbojet and that the propeller blade sweep extends the Mach cruise speed limit for propeller operation.

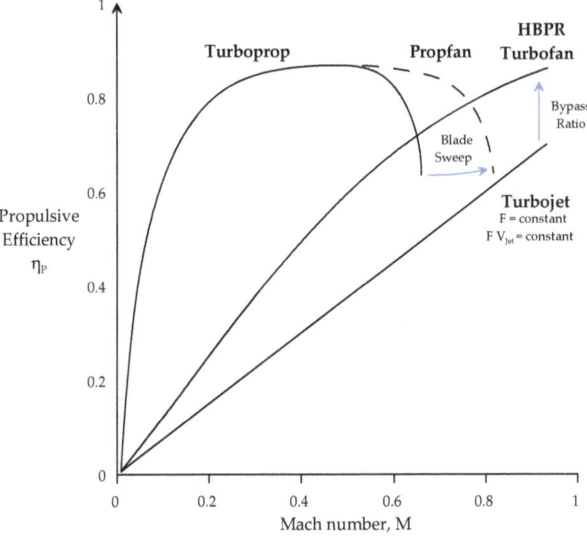

Figure 3. Propulsive efficiency for different engine types.

4. Commercial Aviation—Higher and Faster! Or not?

Although the aviation lemma has typically been higher and faster, in fact, slower commercial aircraft should lead to higher fuel efficiency [26]. Between 1990 and 2010, jet fuel prices have increased over five times, representing about 40 percent of a typical airline's total operating cost [27,28]. As a result, airlines are reviewing all phases of flight to determine how fuel burn savings can be gained in each phase and total.

It is noticeable that since the 1973 oil crisis, commercial airliners are losing their interest in speed from generation to generation (Figure 4). Today, short-haul airliners such as the Boeing 737 and Airbus A320 cruises at airspeeds lower than Mach 0.78. The giant, long-haul flagships like the Boeing 747-8 and the Airbus A380, equipped with four powerful engines, regularly fly at a non-impressing Mach 0.85. Moreover, even the most advanced planes of the modern age as the Boeing 787-Dreamliner and Airbus A350 XWB are not cruising at Mach higher than 0.85. Are the airliners, including the most sublime and advanced of our time, lagging behind the 50-year-old museum exhibits? In the last 50 years, the engines suffered the most noticeable change in aviation. Since the early jet age, engines' bypass ratios increased from 0 to 12.5 (Figure 5). This is easily explained through Equation (3) as engine manufacturers try to increase the engines' propulsive stream mass flow. Modern materials and processes also allowed for higher turbine inlet temperatures and overall pressure ratios. The increase in bypass ratios has been achieved by using bigger fans and fan ducts, which also led to increased empty-weight and parasitic drag. Those factors, associated with rising fuel prices and environmental concerns, are making higher cruise speeds less attractive. In [26], Torenbeek states that "Future long-range airliners optimized for environmentally friendly operation may cruise at no more than Mach 0.75".

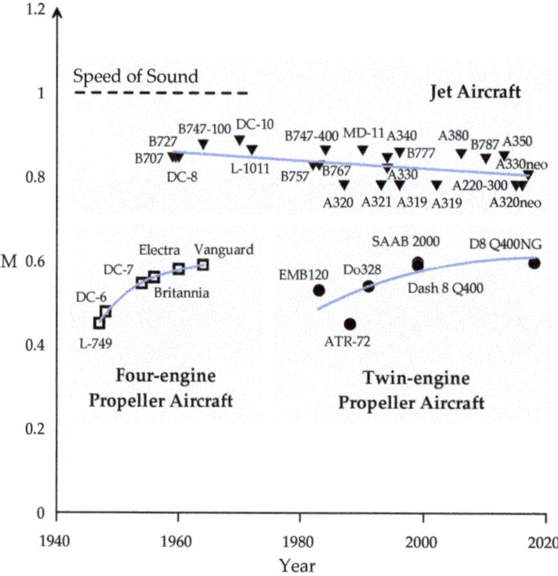

Figure 4. Historical development in maximum cruise speeds for commercial aircraft.

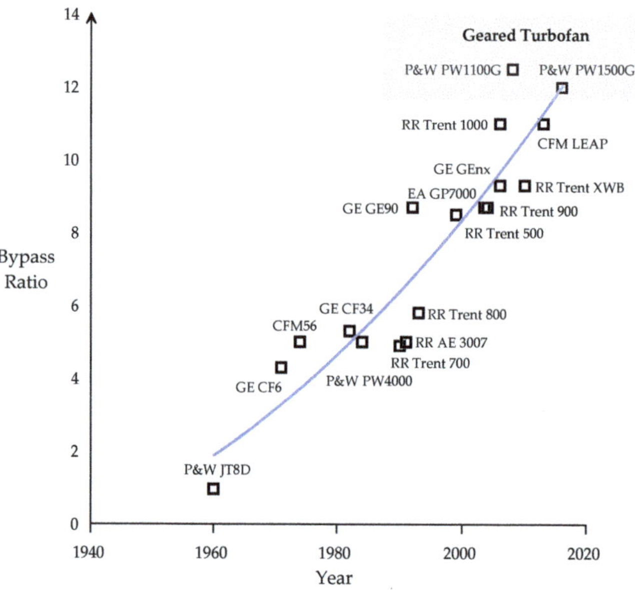

Figure 5. Evolution on the Turbofan Bypass Ratio.

Another noticeable characteristic that confirms this trend of reduced interest in high cruise speeds is the wing design. A comparison of two airliners that operate in the same market segment is presented in Table 1. The wing of Boeing 727 is smaller and has a higher sweep angle than the 737 Max 7. The smaller wingspan and higher wing sweep allow the aircraft to fly faster. When flying near the speed of sound, the airflow accelerates over the wing reaching supersonic speeds and slows down again to subsonic speeds towards the trailing edge of the wing, creating a shockwave and the resulting wave drag. Higher sweep angles delay this effect. The airflow over the wing consists of two components: chordwise flow (parallel to the chord line) and spanwise flow (perpendicular to the chord line). As the only component that suffers the acceleration is the chordwise flow, by reducing the amount of flow in this direction, the aircraft is able to fly faster for the same drag (see Figure 2). Like the newer Boeing 737 Max 7, many other modern airliners have lower wing sweep than their predecessors. This may also be, in part, due to the invention of the supercritical airfoil. However, if the intention were to fly faster, the wing would be kept with a lower span and higher sweep.

Table 1. Boeing 727-200 vs. Boeing 737 Max 7: Technical Specifications. Data from [29,30].

Specifications	Boeing 727-200	Boeing 737 MAX 7
Manufacturing year	1962	2016
Engines	(3×) P&W JT8D-17R	(2×) CFM LEAP-1B
Fuselage length (m)	46.68	35.56
Wingspan (m)	32.92	35.92
Wing sweep	32°	25.03°
MTOW (kg)	95,100	80,286
MMo (Ma)	0.9	0.82
Range–MTOW (Km)	4509	7130
Max Ceiling (m)	13,000	12,000
Fuel capacity (L)	30,620	25,816
Capacity (seats)	155	172

Lower speeds require smaller engine thrust, which represents lower fuel consumption, weight, and noise emissions. Improved efficiency not only allows airlines to save money on fuel but also enables airplanes to fly further. Comparing between the Boeing 727-200 and the 737 Max 7 (described in Table 1), both have similar payload capacities and mass, but the 727-200 has a range of 4500 km, while the 737 Max 7 has 7100 km, twice the range of the former one. In terms of power-plants, the 737 has two engines, while the 727 needed three. The improved take-off thrust of the high bypass turbofan engines was the key to the birth of wide-body airliners, which are the main element of global travel today. However, these high bypass turbofan engines have their drawbacks. All aircraft manufacturers face the same difficulty when upgrading their power-plants. These engines are huge. As the engine manufacturers are raising bypass ratios, the engines' diameter is getting larger, making them harder to fit under the wing of aircraft. The Pratt & Whitney JT8D installed on the 727-200 is much smaller than the CFM Leap-1B that equips the 737 Max 7 (Figure 6). The choice of the Leap-1B to equip the 737 Max 7 also required major upgrades to the landing gear, to maintain the required ground clearance and changes to the wing in order to compensate for the additional engine's weight and drag [31].

The problem is that to adapt the jet propulsion to operate efficiently at lower speeds, the propulsive jet must also have higher mass flow, as observed through Equation (3). Therefore, the turbofan bypass ratio must increase. However, several problems arise when trying to increase the fan size: the fan weight increases; fan noise increases steeply if the peripheral speed is increased to maintain the same shaft rotation speed; reducing the shaft rotational speed such that the noise does not increase, increases the number of stages required for the turbine and thus increases the weight of the core gas turbine. There are a lot more pros than cons to modern aircraft compared to the older airliners. It is a fact that they fly slower, but the rest of their performance is much better, not only due to modern technology but also because of such compromises. In Europe, according to [32], short-haul flights (up to 1500 km) within European Civil Aviation Conference (ECAC) bordering countries represented 78.5% of the total instrument flight rules (IFR) traffic in 2017. According to the same reference, in the United States, the share of short-haul flights reached 80.3% in the same period. Even with the technology that allows us to produce faster airplanes, at those distances, a small increase in speed may reduce flight time but has little impact on the journey. The journey is the wait at the airport, check-in, baggage check, passport control, waiting at the terminal, flight, and again the passport control, baggage claim, and the way from the airport to the destination. All these journey stages will not be accelerated by a higher cruise speed, and all the advantages of flight speed can easily be lost by a traffic jam on the way to the airport. For airlines, the parameters of fuel consumption and life cycle of the aircraft are getting more critical than the cruise speed by the day. Also, fuel consumption is not only money. Fuel tanks on the aircraft remain the same, and flying faster may result in a reduction of range. It is cheaper for the airline to make the passenger more comfortable, show a couple of movies, or provide an extra meal in flight than to speed up the aircraft. From the passenger's point of view, such a deal is also attractive. The flight may be longer, but the level of comfort on those flights is not bad. Higher costs for speed will increase the cost of air tickets, and time is a more valuable resource than money just for a small group of people. The world is ruled by economically optimal airliners with economically optimal performance. A cheaper ticket is more important for a passenger and cheaper operation is more important for airlines. Modern airplanes pursue these goals.

Figure 6. Boeing 727-200 (**left**) side-by-side to a Boeing 737 Max 7 (**right**).

4.1. Present-Day Airliner Speeds

To understand how the typical cruise speeds from the manufacturer compares to the real flight speeds currently being used by airlines, a total of 80 flights were analyzed using Flightradar24. Flightradar24 is a flight tracking service that provides both real-time and stored information about aircraft flights around the world. Specific information such as atmosphere corrected Mach cruise speeds were used to perform this study. Two specific aircraft types were selected: the Boeing 737-800 and the Airbus A320neo. Both aircraft operate short- and medium-haul flights with similar cruise design speeds (with a design Mach cruise speed of 0.785). To compensate for different flight strategies of specific airlines (e.g., low-cost vs full-service), distinct airlines were analyzed for each aircraft type. In Table 2 a synopsis of the analysis is presented.

Table 2. Aircraft analyzed using Flightradar24.

Specifications	Boeing 737-800	Boeing 737-800	Airbus A320neo	Airbus A320neo
Airline	Ryanair	KLM	Wizz Air	British Airways
Registration	9H-QAA	PH-BCK	G-WUKE	G-EUYY
Introduced in	1997	1997	2010	2010
Year of manufacture	2017	2019	2018	2014
V_{Cruise} (Ma) [1]	0.785	0.785	0.78	0.78
V_{Max} (Ma)	0.82	0.82	0.82	0.82
Flights analyzed	20	20	20	20

[1] Design cruise speed announced by the manufacturer.

In Figure 7 the cruise speeds for the total analyzed 80 flights are plotted against aircraft manufacturer announced design cruise speed. From the analyzed flights, it is noticeable that the actual average cruise speed values are lower than the design cruise speed. Regarding the Boeing 737-800, Ryanair presented an average Mach of 0.758 and KLM 0.780. Furthermore, on the Airbus A320neo, Wizz Air average Mach was 0.775, while British Airways registered 0.761. The average Mach cruise speed for the total 80 presented flights is 0.769. Different dispersion can be attributed to distinct weather-related flight optimization, e.g., to account for head or tailwind. Parameters such as airport sockets, aircraft availability, and fuel prices influence each airline's fuel conservation strategies and cost index (CI) . By definition, cost index is the ratio between the time-related operational cost and the fuel cost of an airplane operation, reflecting the relative effects of fuel and time-related operating cost on overall trip cost.

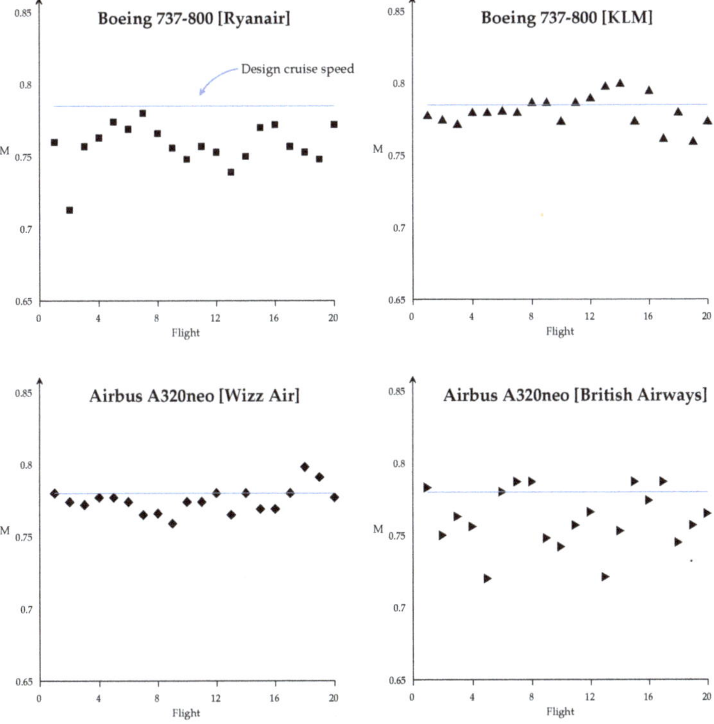

Figure 7. Cruise flight speed from a total of 80 analyzed flights. Boeing 737-800 by Ryanair and KLM. Airbus A320neo by Wizz Air and British Airways. Data sourced from Flightradar24 [33].

From the exclusively propulsive efficiency point of view, referring to Figure 3, one can easily notice that for Mach values of 0.769 the aircraft may already be cruising in a condition where propeller-based propulsive systems, namely propfans, show competitive performance when compared to turbofans [34–36].

5. Propellers Avoidance Phenomenon and the Oil Price Effect

During the jet age, propeller specialists and companies struggled for their place in the industry. After a period of uncertainty, they found it with the development of the turboprop. Since they first emerged in the mid-1940s, turboprop engines were perceived as a temporary compromise between old piston engines and advanced jet engines (see Figure 4, bottom left). This fact resulted in scarce efforts towards technological developments in this type of power-plants and the few turboprop aircraft flying in the late 1960s were still the same built in the 1950s. Though considered rather obsolete, the industry, not seeing great prospects, was not particularly in a hurry to create a replacement for them. In the early 1970s, the use of propeller propulsion in large airframes was almost restricted to the military. However, in 1973, a severe oil crisis [37] started to affect the whole aviation industry. High fuel consumption of the jet engines previously perceived as a perfectly acceptable compromise for speed, associated with prohibitive fuel prices (Figure 8), turned out to be a severe problem. Long-range transportation by large aircraft remained profitable, but flights over short distances by regional vehicles were not often paying off [38]. For the first time in decades, the jet propulsion dominance was questioned, resulting in the industry and governments to chase for more efficient, alternative propulsive systems. This economic environment stimulated work towards a reinvention of the propeller for increased air transportation fuel efficiency.

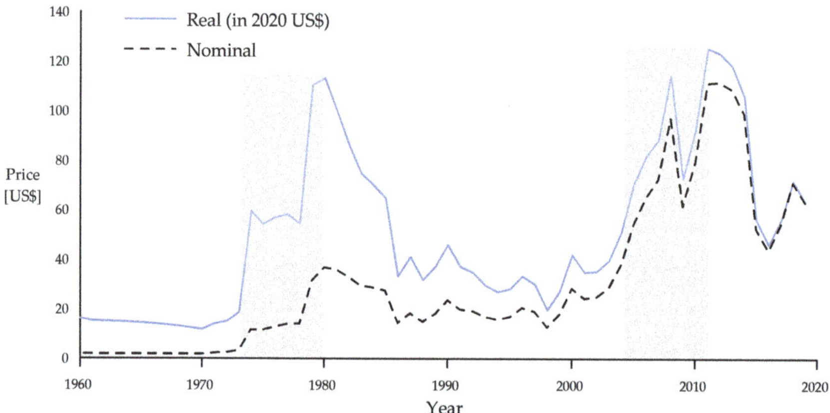

Figure 8. Crude Oil Prices from 1960 to 2020, nominal and real (corrected by the 2019 U.S. inflation). 1960–1985: Arabian Light posted at Ras Tanura; 1986–2020: Brent Spot. Price data source: U.S. Energy Information Administration [39]. U.S. Consumer Price Index (CPI) to correct the prices for the 2019 inflation sourced from the U.S. Bureau of Labor Statistics [40].

5.1. The Advanced Turboprop Project

One of the most remarkable works for increased air transportation fuel efficiency was the Advanced Turboprop Project (ATP) [36]. The ATP was led by the National Aeronautics and Space Administration (NASA) in the 1980s and represented the most relevant and promising work in propeller propulsion up to date. NASA partnering with Boeing and General Electric developed a modern and advanced propeller propulsion system that demonstrated high efficient cruise at Mach 0.65 to 0.80, leading to an overall fuel consumption reduction of 40 to 50 percent relative to turbofans at the time. The flight tests were conducted in a modified B727-100 and McDonnell Douglas MD-80. Later, Pratt & Whitney, cooperating with Allison, developed an even more efficient geared propfan unit that was tested in an MD-80 (Figure 9). In the 1980s, the Advanced Turboprop Project was developing rapidly, and several new aircraft concepts were being considered for the 1990s. Some of them were engine replacements in current aircraft, while others were new aircraft designs specifically for turboprop/propfan installations.

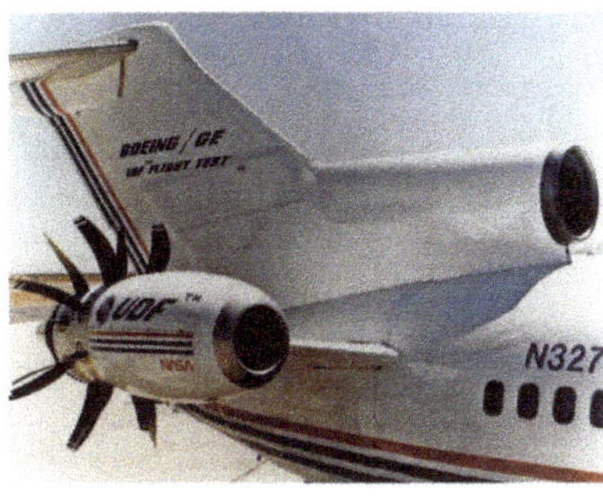

(a)   (b)

Figure 9. (a) Pratt & Whitney-Allison 578–DX geared propfan demonstrator engine, installed on an MD-80 testbed aircraft. (b) General Electric GE36 demonstrator engine installed on the Boeing 727 testbed for flight testing in 1986–1987.

In the meantime, new regional twin-engine turboprop aircraft started to appear to compete with the jets that were becoming too expensive to operate, and the interest for turboprop engines rose again (see Figure 4, bottom right). However, as those new turboprop aircraft started to claim the regional routes, a popular resistance, related to the idea of going back to propeller-driven airplanes, and what it represented in terms of modernity and security, started a propeller avoidance phenomenon on the travelers [3]. This negative perception affected the demand for these routes and impacted the economic viability of the turboprop operated routes. At the same time, a sharp drop in fuel prices (see Figure 8) put down all the research efforts in new and advanced propeller design programs like ATP. The turboprop market decreased, and the competition increased.

Jet planes were more expensive, consumed more fuel, and were more demanding on infrastructure but had better flight performance in factors such as cruise speed, range, and comfort, making them more attractive to operators. This fact led to the lowest demand for turboprop airplanes in the civil transportation market [41] making all those efforts to bring the propeller back to medium/long-haul airliners never materialize.

On the other hand, since its appearance, turboprop aircraft conquered the military and defense businesses. Models like the Lockheed C-130 Hercules and the Lockheed P-3 Orion, introduced in the late 1950s, counting with different variants and updated versions, remain in service up to the present. Their versatile airframe and unprecedented capability to use unprepared runways for takeoffs and landings, made those tactical airlifters to spread out among many military forces worldwide. In the 1970s, Lockheed proposed a C-130 variant with turbofan engines rather than turboprops, but the U.S. Air Force preferred the take-off performance of the existing aircraft [42]. Those characteristics associated with all the accumulated military operational experience and improved performance are demanding for the aircraft industry to come with civil, commercial variants. The Lockheed Martin's LM-100J is a civil derivative of the C-130J Super Hercules, the last major update to the military C-130 family. The LM-100J commercial freighter received its type certificate from the Federal Aviation Administration by the end of 2019, and it is expected to enter service by 2020. To answer the demand for the military tactical airlift and compete with C-130J Super Hercules, Airbus introduced the A400M Atlas in 2009. Equipped with four Europrop TP400-D6 turboprop engines, it has a maximum payload capacity of 37 tons, positioning itself between the C-130J Super Hercules and the larger turbofan Boeing C-17 Globemaster.

The growing concern on fossil fuel outage, combined with the global environmental strategy is increasing the oil price, forcing a paradigm shift in the whole transportation industry. New modern, propeller-powered, hybrid aircraft concepts are emerging. The usage of partial electric power-plants introduces part of the solution to one of the most significant handicaps of propellers in the last decades, the increased ground noise levels. Recently, the Electric Aviation Group (EAG) unveiled the HERA concept (see Figure 10), a Hybrid Electric Regional Aircraft with capacity for 70+ seats to be in service by 2028.

Figure 10. EAG Hybrid Electric Regional Aircraft (HERA) (Photo:PRNewsfoto/EAG).

Reinforcing the prominent comeback of the propellers, in 2008, within the Clean Sky program, the European Commission announced the Open Rotor demonstration program, targeting to reduce fuel consumption and associated CO_2 emissions by 30% compared with current turbofans. Led by Safran (former Snecma), this program assembled a demonstrator in 2015 (see Figure 11), which performed the ground-testing in 2017 at the Istres site in southern France [43].

Figure 11. Open Rotor prototype by Safran.

In response, other companies, like Rolls-Royce, have also been resuming the work and progress from the 1980s ATP program to continue the development and testing on Open Rotor engines, recognizing a clear market opportunity for such technologies in the near future [44].

6. Is There a Future for Aircraft Propellers?

As seen in Section 3, on the one hand, the trend of increasing the bypass ratio in power-plants has been the way to increase the aircraft energetic and economic viability. It has ended up by introducing the gearbox turbofan (see Figure 5). On the other hand, the propeller (or propfan/open rotor fan) can be seen as an ultra-high bypass ratio propulsive device. The gearbox was perceived as a serious disadvantage that disappeared in relation to the recently introduced, geared turbofan. Nevertheless, other disadvantages concerning the use of propellers need to be addressed: higher noise levels and the maximum cruise speed limitation. In Section 4 is shown the acceptance in reducing the maximum cruise speed on consecutive turbofan airliners. On July 16, 2020, Israir, a fairly small Israeli airline with just seven planes in its fleet, including four Airbus A320 and three ATR 72-500, operated a flight from Tel Aviv to Kiev with one of its turboprop ATR 72-500 instead of the A320 (turbofan) [45]. Curiously, this route is stated at 1282 miles, considerably longer than the ATR 72-500 published range of 823 miles, which suggests a low flight load and velocity. The ATR 72-500 performed the flight in 4 h 55 min, while the A320 is capable of 3 h 25 min. Nevertheless, this shows that there are companies already willing to sacrifice flight speed by replacing turbofan aircraft with smaller, more economically viable, propeller-powered aircraft on considerably longer routes.

As stated in Section 5, since the 1970s, several factors were conditioning the broader use of propellers in the civil, commercial aircraft industry. However, the military never dropped the use of propeller aircraft. Their developments in the last decades, associated with current oil prices and ecology strategies, are re-introducing the turboprop aircraft in the civil market.

Finally, the recent developments in future hybrid and electric aircraft brought new challenges to the aircraft designer. When it comes to electric mobility, one of its most significant handicaps is the low energy density of current batteries. Although no disruptive progress is made in that area, the best approach to mitigate that obstacle is through better usage of the limited amount of energy that can be stored. This leads the design approaches back to propeller propulsive systems due to its, more efficient, ultra-high bypass ratios [46].

6.1. Multirotor Drone Emergence and the Future Personal Aerial Mobility

Multirotor drone emergence also re-introduced the propeller as the best-suited propulsion device for low-speed, low-cost, and accessible small aircraft. Furthermore, the enthusiasm for unmanned vehicles, alongside the broader access to technology, are accelerating the interest for personal aerial mobility and package delivery. The present reality of exploding numbers of electric multi rotary-wing drones is being followed by incorporating the fixed-wing flying mode into Vertical Take-off and Landing (VTOL) aircraft for range increase. It is likely that in the near future, electric VTOL (eVTOL) fixed-wing propeller aircraft will serve Uber-like personal mobility transportation systems. As a response to the rising thin-haul and on-demand transportation market, a growing number of startups, and more traditional companies like Airbus and Rolls-Royce are introducing smaller, mostly electric VTOL aircraft for urban air mobility applications. In [47], a review of the current technology and research in urban on-demand air mobility applications, including a comparison between 45 aircraft models, is presented. It is noticeable that in these categories, the propellers and jet propulsion still compete . However, the choice for the propeller-based propulsive systems is taking the lead. The higher efficiency that jet-powered ducted fans offer for high speed, high altitude cruise is not likely to be beneficial for urban applications, where lower speeds and altitudes are more common. In terms of safety, the usage of distributed propulsion is present in almost every new design, allowing improved propulsive efficiency and redundancy. The distributed propulsion takes advantage of typically smaller propellers, also reducing the overall noise with lower tip speeds. Figure 12 shows some of the relevant urban mobility contributions.

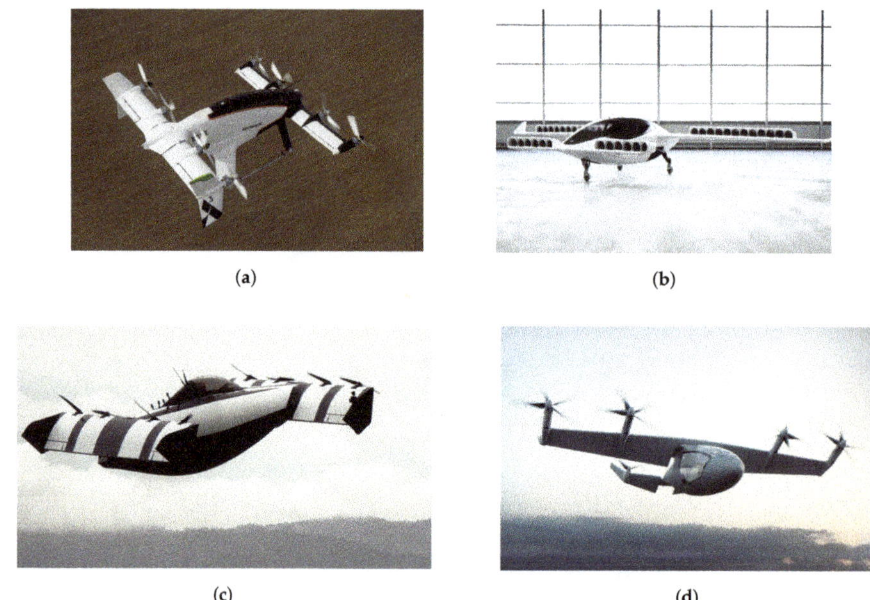

Figure 12. Examples of VTOL aircraft demonstrator vehicles for personal transportation. (**a**) Airbus Vahana, by Airbus Urban Mobility. (**b**) Lilium Jet, by Lilium GmbH. (**c**) BlackFly, by Opener. (**d**) eVTOL, by Rolls-Royce.

7. Conclusions

Since the jet age, aircraft design aimed for speed and drove the airliners cruise speeds up to Mach 0.85. However, the 21st century brought a new sharp rise in energy prices [48], and consequently, a global economic crisis. At the same time, the global environmental consciousness is forcing the reduction of engine emissions, driving recent investigations to suggest that future airliners may have to reduce their design cruise Mach number [26]. Jet regional aircraft are, once again, becoming too expensive to operate, and the demand for turboprop engines rose again. Through the previous energetic crisis, the oil price has driven the progress and technological advance on propellers and their demise in favor of jets, feeding the race to fly higher and faster for decades. The race for speed reached its end, but the efforts to bring the propeller back to medium/long-haul airliners were abandoned as the oil prices dropped. Presently, the higher the oil price, the more likely the propeller comeback is becoming. The industry has played a more reactive than active role in this area. Nevertheless, beyond efficiency, propellers always offered benefits that the jet engine could not. Better take-off and landing performance allow transporting passengers to and from small regional airports. Also, its lower cost allowed enthusiasts and aviators to use them in their recreational, general aviation aircraft. In addition, the demand for turboprop aircraft increased due to a new wave of rising fuel prices, especially in countries that do not have a developed airfield infrastructure. At the very beginning of the 21st century, the propeller-driven airplane prevailed as a niche. However, this century brought new challenges and priorities. Recurrent oil crises are making propellers, and their specific advantages such as economic and environmental, the old answer to a prevailing problem. Therefore, propellers may have a role to play in progress again, even for commercial aviation. Nevertheless, the electric VTOL aircraft aimed for urban mobility seems like the new playground for propeller innovation. As noticed in the previous sections, the propeller has a future and we may be experiencing its rebirth in commercial aviation.

Author Contributions: All authors have contributed to this paper. Conceptualization, P.A. and M.S.; methodology, P.A. and M.S.; formal analysis, P.A. and M.S.; investigation, P.A.; writing—original draft preparation, P.A.;

writing—review and editing, P.A., M.S. and P.G.; visualization, P.A., M.S. and P.G.; supervision, M.S. and P.G.; funding acquisition, M.S. All authors have read and agreed to the published version of the manuscript.

Funding: This work has been supported by the project Centro-01-0145-FEDER-000017 - EMaDeS - Energy, Materials and Sustainable Development, co-financed by the Portugal 2020 Program (PT2020), within the Regional Operational Program of the Center(CENTRO 2020) and the European Union through the European Regional Development Fund (ERDF); and also the C-MAST - UID/EMS/00151 provided by FCT/MCTES through national funds (PIDDAC) and co-financed by the (European Regional Development Fund ERDF) through the Competitiveness and Internationalization Operational Program (POCI).

Conflicts of Interest: The authors declare no conflict of interest.

Abbreviations

The following abbreviations are used in this manuscript:

ATP	Advanced Turboprop Project
CI	Cost Index
CPI	Consumer Price Index
EAG	Electric Aviation Group
ECAC	European Civil Aviation Conference
eVTOL	electric Vertical Take-off and Landing
IFR	Instruments Flight Rules
MMo	Mach Maximum Operating
MTOW	Maximum Take-off Weight
NASA	National Aeronautics and Space Administration
TSFC	Thrust Specific Fuel Consumption
VTOL	Vertical Take-off and Landing
WWI	World War I
WWII	World War II

References

1. Rosen, G.; Anezis, C.A. *Thrusting Forward: A History of the Propeller*, 1st ed.; Hamilton Standard and British Aerospace Dynamics Group: Windsor Locks, CT, USA, 1984.
2. Laufer, B. The Prehistory of Aviation. The Open Court. 1931. Issue 8, Article 5. Available online: https://opensiuc.lib.siu.edu/ocj/vol1931/iss8/5 (accessed on 1 June 2020).
3. Kinney, J.R. *Reinventing the Propeller: Aeronautical Specialty and the Triumph of the Modern Airplane*, 1st ed.; Cambridge University Press: Cambridge, UK, 2017. ISBN 978-1-107-14286-2.
4. *World Intellectual Property Report: Breakthrough Innovation and Economic Growth*; Technical Report; World Intellectual Property Organization: Geneva, Switzerland, 2015. ISBN 978-92-805-2680-6. Available online: https://www.wipo.int/edocs/pubdocs/en/wipo_pub_944_2015 (accessed on 1 June 2020).
5. Douglas, D.W. The Development and Reliability of the Modern Multi-Engine Air Liner. *J. R. Aeronaut. Soc.* **1935**, *39*, 1010–1046. also reprinted in *J. Aeronaut. Sci.* **1935**, *2*, 128–152, doi:10.2514/8.99. [CrossRef]
6. Valavanis, K.P. (Ed.) *Advances in Unmanned Aerial Vehicles: State of the Art and the Road to Autonomy*; Springer: Tampa, FL, USA, 2007. ISBN 978-1-4020-6113-4.
7. Dumas, A.; Trancossi, M.; Madonia, M.; Giuliani, I. Multibody Advanced Airship for Transport. In Proceedings of the Aerospace Technology Conference and Exposition, SAE International, Warrendale, PA, USA, 18 October 2011. doi:10.4271/2011-01-2786. [CrossRef]
8. Martin, P.; Devasia, S.; Paden, B. A different look at output tracking: control of a vtol aircraft. *Automatica* **1996**, *32*, 101–107. doi:10.1016/0005-1098(95)00099-2. [CrossRef]
9. Erginer, B.; Altug, E. Modeling and PD Control of a Quadrotor VTOL Vehicle. In Proceedings of the 2007 IEEE Intelligent Vehicles Symposium, Istanbul, Turkey, 13–15 June 2007; pp. 894–899.
10. Olfati-Saber, R. Global configuration stabilization for the VTOL aircraft with strong input coupling. *IEEE Trans. Autom. Control* **2002**, *47*, 1949–1951. doi:10.1109/TAC.2002.804457. [CrossRef]
11. Herisse, B.; Russotto, F.X.; Hamel, T.; Mahony, R. Hovering flight and vertical landing control of a VTOL Unmanned Aerial Vehicle using optical flow. In Proceedings of the 2008 IEEE/RSJ International Conference on Intelligent Robots and Systems (IROS), Nice, France, 22–26 September 2008; pp. 801–806. doi:10.1109/IROS.2008.4650731. [CrossRef]

12. ACARE. (Ed.) Strategic Research Agenda; 2002; Volume 1. Available online: https://www.acare4europe.org/sites/acare4europe.org/files/document/volume1.pdf (accessed on 1 June 2020).

13. Verhovek, S.H. *Jet Age: The Comet, the 707, and the Race to Shrink the World*, 1st ed.; Avery: New York, NY, USA, 2010. ISBN 978-15-8333-402-7.

14. Talay, T.A. *Introduction to the Aerodynamics of Flight*; Technical Report SP-367; NASA: Washington, DC, USA, 1975. Available online: https://ntrs.nasa.gov/archive/nasa/casi.ntrs.nasa.gov/19760003955.pdf (accessed on 1 June 2020).

15. Whitford, R. *Design for Air Combat*, 1st ed.; Jane's Information Group: London, UK, 1987. ISBN 9780710604262.

16. Proctor, J. *Convair 880 & 990: Great Airliners Series*, 1st ed.; World Transport Press: 1996; Volume 1. ISBN 9780962673047.

17. Boeing Supersonic Transport: Historical Snapshot Web Page. Available online: https://www.boeing.com/history/products/supersonic-transport.page (accessed on 1 June 2020).

18. Mckee, J.W. Flight Control System for The Boeing 2707 Supersonic Transport Airplane. SAE Technical Paper 670528. In Proceedings of the Aerospace Systems Conference and Engineering Display, SAE International, Troy, MI, USA, 1 February 1967. doi:10.4271/670528. [CrossRef]

19. McLean, F.E. *Supersonic Cuise Technology. NASA SP-472*.; NASA: Washington, DC, USA, 1985. Available online: https://ntrs.nasa.gov/archive/nasa/casi.ntrs.nasa.gov/19850020600.pdf (accessed on 1 June 2020).

20. Hearth, D.P.; Preyss, A.E. Hypersonic technology-approach to an expanded program. *Astronaut. Aeronaut.* **1976**, *14*, 20–37.

21. Hallion, R. The History of Hypersonics: Or, 'Back to the Future: Again and Again'. In Proceedings of the 43rd AIAA Aerospace Sciences Meeting and Exhibit, Reno, NV, USA, 10–13 January 2005. doi:10.2514/6.2005-329. [CrossRef]

22. Rutherford, D.; Graver, B.; Chen, C. Noise and Climate Impacts of an Unconstrained Commercial Supersonic Network. The International Council On CLean Transportation. 2019. Available online: https://theicct.org/publications/noise-climate-impacts-unconstrained-supersonics (accessed on 1 June 2020).

23. Bramson, D. The Concorde: A Supersonic Airplane Too Advanced to Survive. The Atlantic. 1 July 2015. Available online: https://www.theatlantic.com/technology/archive/2015/07/supersonic-airplanes-concorde/396698/ (accessed on 1 June 2020).

24. Farokhi, S. *Aircraft Propulsion*, 2nd ed.; WILEY: Chichester, UK, 2014. ISBN 978-1118806777.

25. Concorde fuel Performance: Better Than It Seems? *FLIGHT International*, 8 May 1976.

26. Torenbeek, E. *Advanced Aircraft Design: Conceptual Design, Analysis and Optimization of Subsonic Civil Airplanes*; John Wiley & Sons: Delft, The Netherlands, 2013. ISBN 978-1-118-56811-8.

27. Schulte, D. Estimating Maintenance Reserves. Boeing Aero, Q4 2013, pp. 5–11. Available online: https://www.boeing.com/commercial/aeromagazine/articles/2013_q4/pdf/AERO_2013q4.pdf (accessed on 1 June 2020).

28. Roberson, W.; Johns, J.A. Fuel Conservation Strategies: Takeoff and Climb. Boeing Aero, Q4 2008, pp. 25–28. Available online: https://www.boeing.com/commercial/aeromagazine/articles/qtr_4_08/pdfs/AERO_Q408.pdf (accessed on 1 June 2020).

29. Boeing 727-200 Technical Specifications. Available online: https://www.airliners.net/aircraft-data/boeing-727-200/90 (accessed on: 1 June 2020).

30. Boeing 737 Max Web Page. Available online: https://www.boeing.com/commercial/737max/ (accessed on 1 June 2020).

31. Ostrower, J. Boeing Narrows 737 Max Engine Fan Size Options to Two. FlightGlobal. 31 August 2011. Available online: https://www.flightglobal.com/news/articles/boeing-narrows-737-max-engine-fan-size-options-to-two-361438/ (accessed on 1 June 2020).

32. EUROCONTROL; FAA. U.S./Europe Comparison of Air Traffic Management-Related Operational Performance for 2017. 8 April 2019. Available online: https://www.eurocontrol.int/publication/useurope-comparison-air-traffic-management-related-operational-performance-2017 (accessed on 1 June 2020).

33. Flightradar24 Web Page. Available online: https://www.flightradar24.com/ (accessed on 1 June 2020).

34. Reynolds, C.N. Advanced Prop-fan Engine Technology (APET) Single- and Counter-Rotation Gearbox/Pitch Change Mechanism; NASA: Cleveland, OH, USA, 1985. Available online: https://ntrs.nasa.gov/archive/nasa/casi.ntrs.nasa.gov/19870019119.pdf (accessed on 1 June 2020).

35. Sargisson, D. *Advanced Propfan Engine Technology (APET) and Single-Rotation Gearbox/Pitch Change Mechanism*; NASA: Cleveland, OH, USA, 1985. Available online: https://ntrs.nasa.gov/archive/nasa/casi.ntrs.nasa.gov/19870019120.pdf (accessed on 1 June 2020).

36. Hager, R.D.; Vrabel, D. *Advanced Turboprop Project*; NASA: Cleveland, OH, USA, 1988. Available online: https://ntrs.nasa.gov/archive/nasa/casi.ntrs.nasa.gov/19890003194.pdf (accessed on 1 Junte 2020).

37. Issawi, C. The 1973 Oil Crisis and After. *J. Post Keynes. Econ.* **1978**, *1*, 3–26, doi:10.1080/01603477.1978.11489099. [CrossRef]

38. Hallion, R.P. (Ed.) *NASA's Contributions to Aeronautics*; NASA: Washington, DC, USA, 2010; Volume 1, p. 1041. ISBN 978-0-16-084635-9. Available online: https://www.nasa.gov/connect/ebooks/aero_contributions1_detail.html (accessed on 1 June 2020).

39. EIA. U.S. Energy Information Administration Web Page. Available online: https://www.eia.gov (accessed on 1 June 2020).

40. United States Department of Labor. Bureau of Labor Statistics Web Page. Available online https://www.bls.gov/cpi/ (accessed on 1 June 2020).

41. von Schoenberg, A. Turboprop Aircraft Insight 2015. Technical Report, The Sharp Wings. 2015. Available online: http://www.thesharpwings.com/wp-content/uploads/2015/12/The-Sharpwings-Turboprop-Aircraft-Insight-20152.pdf (accessed on 1 June 2020).

42. Cawthorne, N. *The Mammoth Book of Inside the Elite Forces*; Running Press: Philadelphia, PA, USA, 2008.

43. Safran Celebrates Successful Start of Open Rotor Demonstrator Tests on New Open-Air Test Rig in Southern France. Available online: https://www.safran-group.com/media/safran-celebrates-successful-start-open-rotor-demonstrator-tests-new-open-air-test-rig-southern-france-20171003 (accessed on 21 July 2020).

44. Rolls-Royce Shares Next Generation Engine Designs. Available online: https://www.rolls-royce.com/media/press-releases/2014/260214-next-generation.aspx (accessed on 21 July 2020).

45. Israir's Unbelievably Long Turboprop Flight. Available online: https://onemileatatime.com/israir-long-turboprop-flight/ (accessed on 21 July 2020).

46. Hoelzen, J.; Liu, Y.; Bensmann, B.; Winnefeld, C.; Elham, A.; Friedrichs, J.; Hanke-Rauschenbach, R. Conceptual Design of Operation Strategies for Hybrid Electric Aircraft. *Energies* **2018**, *11*, 217. [CrossRef]

47. Polaczyk, N.; Trombino, E.; Wei, P.; Mitici, M. A review of current technology and research in urban on-demand air mobility applications. In Proceedings of the 8th Biennial Autonomous VTOL Technical Meeting and 6th Annual Electric VTOL Symposium 2019, Vertical Flight Society, Mesa, AZ, USA, 28 January–1 February 2019; pp. 333–343. ISBN 9781510888746.

48. Hendricks, R.C.; Daggett, D.L.; Anast, P.; Lowery, N. Future Fuel Scenarios and Their Potential Impact to Aviation. In Proceedings of the 11th International Symposium on Transport Phenomena and Dynamics of Rotating Machinery, Honolulu, HI, USA, 25 February–2 March 2006. Available online: https://ntrs.nasa.gov/archive/nasa/casi.ntrs.nasa.gov/20110012886.pdf (accessed on 1 June 2020).

Article

Proportional Resonant Current Control and Output-Filter Design Optimization for Grid-Tied Inverters Using Grey Wolf Optimizer

João Faria, João Fermeiro, José Pombo, Maria Calado and Sílvio Mariano *

Instituto de Telecomunicações, Universidade da Beira Interior, Covilhã 6201-001, Portugal; joao.pedro.faria@ubi.pt (J.F.); fermeiro@ubi.pt (J.F.); jose.pombo@ubi.pt (J.P.); rc@ubi.pt (M.C.)
* Correspondence: sm@ubi.pt; Tel.: +35-127-532-9760

Received: 4 March 2020; Accepted: 10 April 2020; Published: 14 April 2020

Abstract: This paper proposes a new method for the simultaneous determination of the optimal control parameters of proportional resonant controllers and the optimal design of the output filter of a grid-tied three-phase inverter. The proposed method, based on the grey wolf optimization (GWO) algorithm, addresses both optimization problems as a single process to achieve a better system frequency response. It optimizes the unknown parameters by using a fitness function to find the best trade-off between the following fundamental terms: the harmonic attenuation rate; the power loss, through the damping resistor; and the current tracking error in the stationary frame, ensuring the system and grid stability. To validate the proposed optimization methodology, two case studies are considered with different output filter topologies with passive damping methods. The results obtained from the proposed optimization procedure were analyzed and discussed according to the fitness function terms.

Keywords: grid-tied inverter; grey wolf optimizer; PR controllers; LCL filter; passive damping

---

1. Introduction

The over-exploitation of non-renewable natural resources has led an overall environmental degradation on the planet. A suitable alternative solution to mitigate the environmental degradation on the planet is the use of renewable energy sources [1,2]. Nowadays, solar and wind energy are some of the most attractive renewable energy sources for electrical energy production [3]. However, to correctly accomplish the connection of this type of renewable production to the electrical grid, it is necessary to use a grid-tied inverter [4]. There are some essential factors to ensure that the coupling between the inverter and the electrical grid is carried out efficiently and meets the specified standards [5]. These factors can be roughly divided into three categories: control structures, control parameters optimization, and hardware configuration [6]. Different control structures have been applied and can be divided into linear or non-linear structures. The most commonly used controllers with a nonlinear control structure are hysteresis controllers, predictive controllers, and deadbeat controllers [7]. Hysteresis controllers are widely used because of their characteristics: simple implementation, low cost, and fast dynamic response [8,9]. However, these controllers have the disadvantages of introducing harmonics into the current shape. Furthermore, they operate with a variable switching frequency, making the output filter design more complex [10].

Predictive controllers estimate the future value of the output current based on the previous current value. These controllers provide a more precise control and lower current shape distortion. However, they require a detailed model of the system and its implementation is more complex [11]. In addition, when model plant mismatches are presented, this will influence the control accuracy and a significant performance degradation is observed [12]. Deadbeat controllers provide fast dynamics by forcing the system's response to follow the current or voltage reference in the smallest number of time steps [13]. The deadbeat controllers require a detailed model plant and a higher sampling rate, and are very sensitive to disturbances on the system's inherent parameters.

Regarding linear control structures, the most widely used controllers found in the literature are the proportional integral (PI) controllers and proportional resonant (PR) controllers [7]. PI controllers are quite popular for their stability, efficiency, and good dynamic response to reference power change and unbalanced grid faults conditions [14]. However, when applied to the stationary frame, this type of controllers does not have the ability to follow a sinusoidal reference without a steady-state error [11]. To use this type of controller, it is necessary to transform the system variables into the synchronous frame, which requires a large amount of trigonometric calculations.

To avoid the system variables' transformations to the synchronous frame, PR controllers are used, providing a high gain around the resonance frequency, enforcing a zero steady-state error. Consequently, they present a faster control response, a simpler control structure, and a good capability to compensate harmonics components [15]. Furthermore, it is possible to combine PR controllers with selective harmonic compensators (HCs) to fulfill the constraints established by the IEEE-519 and IEEE-1547 standards. Further, to increase the domain stability, in the case of grid frequency deviations, several modifications to PR controllers have been proposed [7,16]. There are several methods in the literature for designing these controllers [17]. However, few studies address this design as an optimization process (tuning).

Indeed, optimization of control parameters is a fundamental aspect to maximize the efficiency and robustness of the system. Through this optimization process, it is possible to guarantee the stability of the system and, at the same time, improve its dynamic response in terms of overshoot, oscillation, rising, and settling times. Many conventional, statistical, and metaheuristic methods have been used to optimize the controller parameters. However, in recent years, the metaheuristic methods have emerged as a competitive approach owing to their characteristics. Therefore, metaheuristic algorithms, such as Cuckoo search optimization algorithm (CS) [18], grey wolf optimization algorithm (GWO) [19], fruit fly optimization algorithm (FOA) [6], grasshopper optimization algorithm (GOA) [20], bat optimization algorithm (BAT) [21], particle swarm optimization algorithm (PSO) [22], salp swarm optimization algorithm (SSA) [23], and whale optimization algorithm (WOA) [24], have been widely applied to optimize the controller parameters.

Another essential aspect to suppress the high-order harmonics introduced by the pulse-width modulation (PWM) technique used in grid-tied inverters is the output filter design. Different passive output filter topologies are used [25]. The *L* output filters are first-order filters that allow an attenuation of 20 dB/dec. However, this type of output filter has poor performance at higher frequency components. In this way, a large inductance is required to limit the high frequency switching ripple, which results in a bulky and expensive passive filter [26]. By including a capacitance (C), it is possible to obtain a second-order output filter (*LC*) that allows an attenuation of 40 dB/dec. This type of output filter has higher efficiency, lower cost, and smaller dimensions. The *LCL* output filters are third-order filters that allow an attenuation of 60 dB/dec and are a standard solution for grid-tied inverters. When compared with the other two, this type of filter achieves greater attenuation of high frequency components. However, the *LC* and *LCL* output filters have the disadvantage of introducing a resonance frequency in the system, which can cause distortion in the output current shape and, at the worst-case scenario, loss of stability.

To mitigate this disadvantage, several passive and active damping methods have been proposed in the literature. Passive damping methods consist in introducing passive elements in the filter structure and are classified in three groups: series passive damping (SPD), parallel passive damping (PPD), and complex passive damping (CPD) [27]. These damping methods have less control complexity and are more reliable, however, they introduce additional losses. On the other hand, active damping methods are more efficient, but require more complex control, being more selective in their action; typically need extra sensors; and are more sensitive to parameter uncertainties [28]. As a result, designing output filters and the corresponding damping method is a complex task because efficiency and damping effect are antagonistic and interrelated goals. There are several methods in the literature based on analytical solutions, simple approximations, and numerical and metaheuristic methods to obtain the *LCL*-filter configuration fulfilling the grid requirements while incurring minimal energy storage and losses, as well as total harmonic distortion (THD) [29–33]. In [30], the *LCL* filter elements were optimized for minimum energy stored using an analytic solution. In [32], an optimal design of the *LCL* filter based on a analytic solution was proposed. In [31], simple approximations were used to calculate the values of the *LCL* filter elements and minimize the power losses. In [33], a combination of partial direct-pole-placement and differential evolution algorithm was adopted to determine basic parameters of the proportional resonant controller for the grid-tied inverters using the *LCL* filter. In [34], an multiobjective evolutionary algorithm was used to design an *LCL* filter with minimum inductance cost, maximum harmonic attenuation rate, and best current tracking effect.

This paper proposes a new method, based on the GWO optimization algorithm, to determine in a single optimization process the optimal parameters of the PR current controllers and the design of the output filter. The proposed method addresses both optimization problems as a single process, to achieve a complete symbiosis and a better frequency response of the whole system. In addition, the fitness function considered tries to achieve an agreement between the fundamental terms that maximize the system efficiency. The proposed method considers the system and grid stability constraints to maximize the harmonic attenuation rate, minimize the power loss through the damping resistor, and optimize the current tracking error. To validate the proposed methodology, two case studies were considered, with different output filter structures. The first case study was carried out with a series passive damping (SPD) topology, commonly found in the literature, for comparison purposes. The second case study consists in complex passive damping (CPD) topologies, as the design of the output filter is a complex process and there is no optimal paradigm to do that. The simulation environment Matlab/Simulink® was used to carry out the simulations. The results obtained from the proposed optimization procedure are analyzed and discussed.

This paper is organized as follows. Section 2 presents the control schemes used in grid-tied inverters, emphasizing the current controllers and different output filters structures. Section 3 describes the proposed methodology, namely, the output filter topologies used, the implemented current controller, the implemented optimization algorithm, and the conditions of the optimization process. Section 4 presents the results obtained from the proposed optimization procedure. Section 5 concludes the paper and discusses the achieved results.

2. Control Schemes

In order to ensure the correct operation of a grid-tied inverter, a robust and effective control is necessary, which guarantees compliance with all operating standards for this type of converter [5]. However, despite being transversal to all grid-tied inverters, this type of control is highly complex and, therefore, has been studied with great interest by the scientific community. Figure 1 shows the main control schemes that are used in grid-tied inverters. These control scheme can be divided in four groups: in a first group, there is maximum power point tracking (MPPT), which controls the DC-DC converter responsible for extracting the maximum power from endogenous and renewable sources [35]; the second group consists of the voltage control of the DC bus. This controller is responsible for controlling the voltage on the DC bus, ensuring the energy balance of the system at any instant [36]; in a third group, there is a current control that ensures that the current injected to the electric grid follows the specified reference; finally, in a fourth group, there is a control scheme associated to the synchronism with the electrical grid and responsible for extracting its voltage characteristics [37].

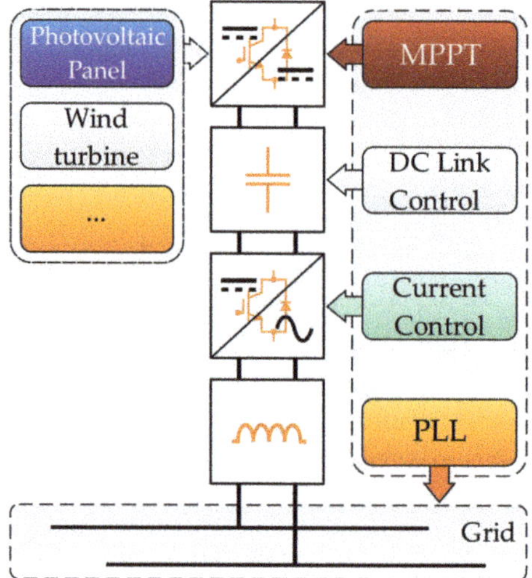

Figure 1. Main control schemes associated with a grid-tied inverter. MPPT, maximum power point tracking; PLL, phase-locked loop.

To ensure compliance with the standards specified in [5], namely the total harmonic distortion (THD), it is essential to optimize the current controller and the design of the output filter. The current controllers and the output filters most widespread in the literature are presented below.

2.1. PR Current Controller

A well-established solution in the literatures is the use of PR controllers, illustrated in Figure 2. These controllers are characterized by only having a high gain in their resonance frequency, allowing it to achieve a zero steady-state error in the stationary frame.

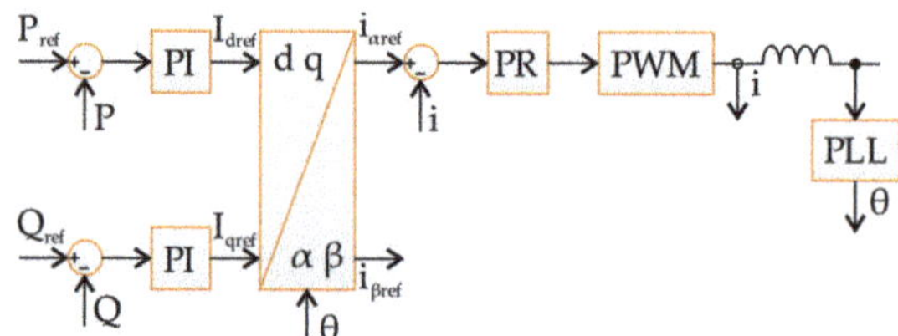

Figure 2. Block diagram of the proportional resonant (PR) current controller. PI, proportional integral; PWM, pulse-width modulation.

The ideal transfer function of PR controllers is expressed by Equation (1), where w_n is the resonant frequency of the controller and the parameters k_p and k_r are the proportional and resonant gains, respectively.

$$H(s) = k_p + k_r \frac{s}{s^2 + w_n^2} \tag{1}$$

In Figure 3, it is possible to see a comparison of the frequency responses of PI and ideal PR controllers. Regarding the PI controllers, it is possible to verify that the gain of the PI controller is inversely proportional to the frequency, losing the ability to cancel the steady state error with increasing frequency [38]. PR controllers have a high gain around the resonance frequency and their bandwidth depends on the value of k_r-a small value creates a very narrow band, while a large value of k_p will cause a larger band around the resonance frequency.

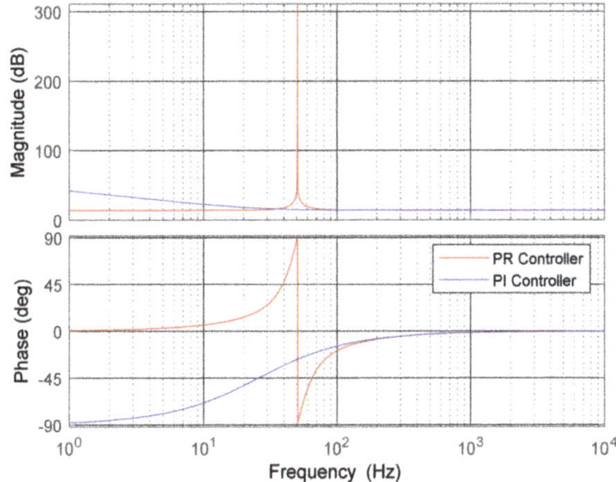

Figure 3. Frequency response of PI and ideal PR controllers, with $k_p = 5$, $k_i = 800$, $k_r = 800$, and $w_n = 214$ rad/s.

To avoid stability problems resulting from the high gain in the resonance frequency, the non-ideal transfer function of the PR controllers is used, expressed by Equation (2). In this transfer function, the expression $w_c \ll w_n$ is assumed, where w_c is the bandwidth around the resonance frequency w_n.

$$H(s) = k_p + \frac{2k_i\left(w_c s + w_n^2\right)}{s^2 + 2w_c s + w_n^2 + w_c^2} \approx k_p + \frac{2k_i w_c s}{s^2 + 2w_c s + w_n^2} \tag{2}$$

Using Tustin's approximation with pre-warped frequency [39] to discretize the system expressed by Equation (2), that is, considering $s = \underbrace{\frac{w_n}{tan\left(\frac{w_n T_s}{2}\right)}}_{k_t} \frac{z-1}{z+1}$, we get the following:

$$H(s) = k_p + \left(\frac{b_0 + b_1 z^{-2}}{a_0 + a_1 z^{-1} + a_2 z^{-2}}\right) \tag{3}$$

where

$$b_0 = 2k_r w_c k_t \tag{4}$$

$$b_1 = -2k_r w_c k_t \tag{5}$$

$$a_0 = K_t^2 + 2w_c k_t + w_n^2 \tag{6}$$

$$a_1 = 2w_n^2 - 2k_t^2 \tag{7}$$

$$a_2 = K_t^2 - 2w_c k_t + w_n^2 \tag{8}$$

In Figure 4, we can see the frequency response of the non-ideal PR controller in the continuous and discrete domains. As can be seen, there is a high bandwidth around the resonance frequency, allowing to increase its stability and robustness in situations of grid frequency deviations. Furthermore, the gain is now finite, but still relatively high for enforcing a zero steady-state error.

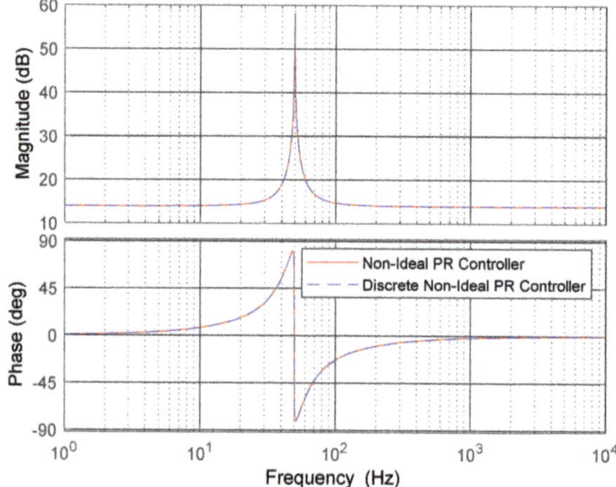

Figure 4. Frequency response of the non-ideal PR controller in the continuous and discrete domains, with $k_p = 5$, $k_r = 800$, $w_n = 314$ rad/s, and $w_c = 10$ rad/s.

2.2. Output Filter Topologies

Several output filter topologies have been documented in the literature to improve the coupling between the inverter and the electric grid, by limiting the current harmonics injected in the point of common coupling (PCC) [40]. The most common output filters are the *L*, *LC*, and *LCL* filters, illustrated in Figure 5.

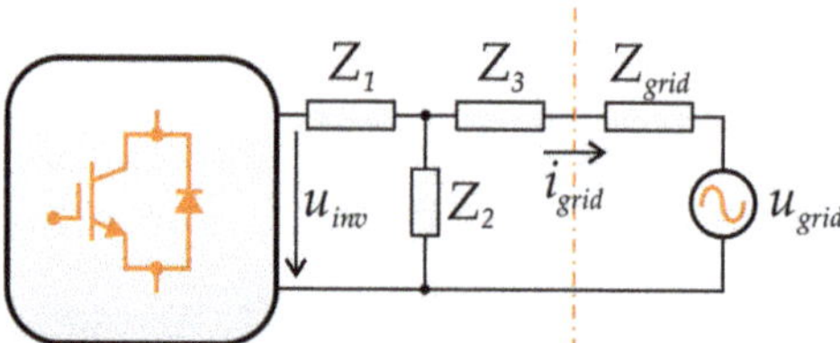

Figure 5. Representation of the output filter structure of the grid-tied inverter.

The output filters *L*, where Z_1 is finite, Z_2 is infinite, and Z_3 equal to zero, are first order filters with an attenuation of 20 dB/dec. The ideal transfer function of this type of filter is given by Equation (9).

$$H(s) = \left. \frac{i_{grid}}{u_{inv}} \right|_{\substack{u_{grid} = 0 \\ z_{grid} = 0}} = \frac{1}{Ls} \tag{9}$$

The output filters (*LC*), where Z_1 is finite, Z_2 is finite, and Z_3 equal to zero, are second-order filters that allow an attenuation of 40 dB/dec. This type of structure exhibits greater performance in the attenuation of high frequency components. However, it introduces a resonant frequency in the system, causing distortions in the output current shape or loss of stability. The ideal transfer function of this type of filter is expressed by Equation (10).

$$H(s) = \left. \frac{u_{grid}}{u_{inv}} \right|_{z_{grid}=0} = \frac{1}{s^2 LC + 1} \tag{10}$$

The *LCL* filters, considering Z_1, Z_2, and Z_3 are finite, are third order filters that allow an attenuation of 60 dB/dec. This type of structure offers greater attenuation of the high frequency components when compared with the filters previously described. The ideal transfer function of *LCL* filters is given by Equation (11).

$$H(s) = \left. \frac{i_{grid}}{u_{inv}} \right|_{\substack{u_{grid}=0 \\ z_{grid}=0}} = \frac{1}{s^3 L_1 L_2 C + s(L_1 + L_2)} \tag{11}$$

Figure 6 shows six passive damping topologies that can be classified into three groups: series passive damping (SPD); parallel passive damping (PPD), and complex passive damping (CPD) [27]. Figure 6a shows an SPD topology commonly found in the literature, consisting of the introduction of an R_d resistance in the capacitor branch. The PPD topology, illustrated in Figure 6b,c, consists of introducing a resistance in parallel in the structure of the output filter in order to have a proper damping effect and stability. Figure 6d–f illustrate three CPD topologies that consist of introducing an impedance path (composed by resistors, inductors and capacitors) in parallel/series with the capacitor. With this type of topology, it is possible to reduce the power loss in the damping resistor at the fundamental and high order harmonics, and thus achieve greater efficiency. However, its design is a more complex process and there is no optimal paradigm to determine the filter elements.

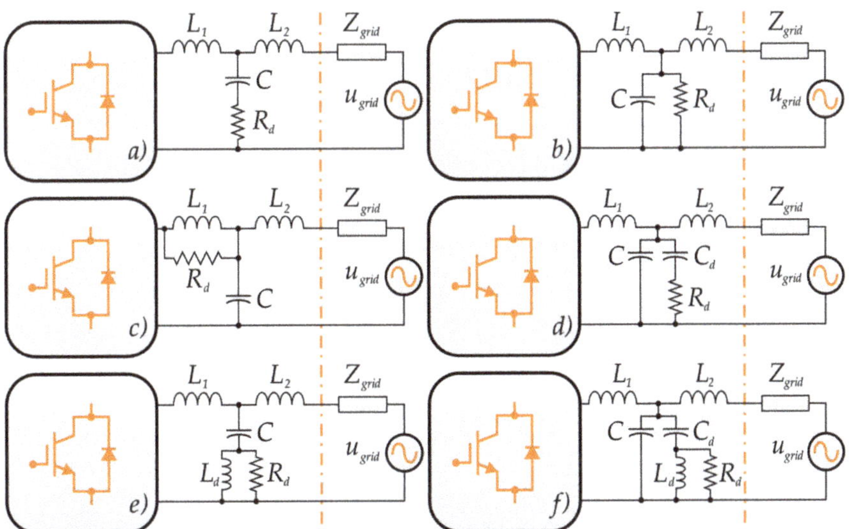

Figure 6. Output filters topologies with passive damping. (**a**) SPD topology commonly found in the literature; (**b**,**c**) consists of introducing a resistance in parallel in the structure of the output filter in order to have a proper damping effect and stability; (**d**–**f**) three CPD topologies that consist of introducing an impedance path (composed by resistors, inductors and capacitors) in parallel/series with the capacitor.

3. Optimization Procedure

To determine the optimum parameters of the PR controllers and the design of the output filter, a single optimization procedure was used. Although this optimization is a more complex computational task, it provides a better frequency response of the system because both frequency responses are correlated and their performances are mutually affected. To validate the proposed methodology, two case studies with four different structures of output filters were considered, as shown in Figure 7. The first case study consists of an SPD topology commonly found in the literature, for comparison purposes. Therefore, the optimization problem with topology 1 consists in determining the optimal values for the six unknown parameters $\tau = [L_1, L_2, C, R_d, k_r, k_p]$. As the design of the output filter is a complex process and there is no optimal paradigm, three CPD topologies were used in the second case study. The optimization problem with topology 2 involves the determination of optimal values for the seven unknown parameters $\tau = \left[L_1, L_2, C, C_d, R_d, k_r, k_p\right]$. For topologies 3 and 4, the optimization problem consists in determining the seventh and eighth unknown parameters $\tau = \left[L_1, L_2, C, L_d, R_d, k_r, k_p\right]$ and $\tau = \left[L_1, L_2, C, C_d, L_d, R_d, k_r, k_p\right]$, respectively. For simulation purposes, the system illustrated in Figure 7 was implemented, consisting of a three-phase three-wire inverter topology with an SPWM with a switching frequency of 20 kHz.

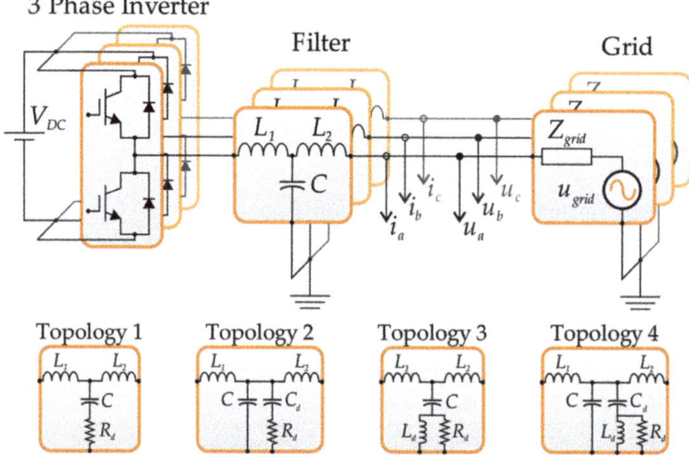

Figure 7. System representation with the output filter topologies used.

Equations (12)–(15) show the transfer function that relates the current grid (i_{grid}) with the converter voltage (u_{inv}) of the four topologies, used to validate the proposed methodology.

$$H(s) = \left. \frac{i_{grid}}{u_{inv}} \right|_{\substack{u_{grid}=0 \\ z_{grid}=0}} = \frac{R_d C s + 1}{s^3 L_1 L_2 C + s^2 R_d C (L_1 + L_2) + s(L_1 + L_2)} \tag{12}$$

$$H(s) = \left. \frac{i_{grid}}{u_{inv}} \right|_{u_{grid}=0} = \frac{R_d C_d s + 1}{s^4 L_1 L_2 C C_d R_d + s^3 L_1 L_2 (C + C_d) + s^2 R_d C_d (L_1 + L_2) + s(L_1 + L_2)} \tag{13}$$

$$H(s) = \left. \frac{i_{grid}}{u_{inv}} \right|_{\substack{u_{grid}=0 \\ z_{grid}=0}} = \frac{s^2 L_d R_d C + s L_d + R_d}{s^4 L_1 L_2 L_d C + s^3 R_d C (L_1 L_2 + L_1 L_d + L_2 L_d) + s^2 (L_1 L_d + L_2 L_d) + s R_d (L_1 + L_2)} \tag{14}$$

$$H(s) = \left. \frac{i_{grid}}{u_{inv}} \right|_{\substack{u_{grid}=0 \\ z_{grid}=0}} = \frac{s^2 L_d R_d C_d + s L_d + R_d}{\substack{s^5 L_1 L_2 L_d C C_d R_d + s^4 L_1 L_2 L_d C + s^3 R_d (L_1 L_2 (C + C_d) + L_1 L_d C_d + L_2 L_d C_d) + \ldots \\ + s^2 (L_1 L_2 L_d C + L_1 L_d + L_2 L_d) + s R_d (L_1 + L_2)}} \tag{15}$$

The implemented control structure is presented in Figure 8. PR controllers were used in both axes of the stationary frame (α and β) and implemented through Equations (3)–(8). The synchronization with the grid is performed using the second order generalized integrator phase-locked loop (SOGI-PLL) algorithm, characterized by Equations (16) and (17) with a sampling frequency of 20 kHz.

$$H_d(s) = \frac{\widehat{u_d}}{u_{grid}} = \frac{k w_n s}{s^2 + k w_n s + w_n^2} \tag{16}$$

$$H_q(s) = \frac{\widehat{u_q}}{u_{grid}} = \frac{k w_n^2}{s^2 + k w_n s + w_n^2} \tag{17}$$

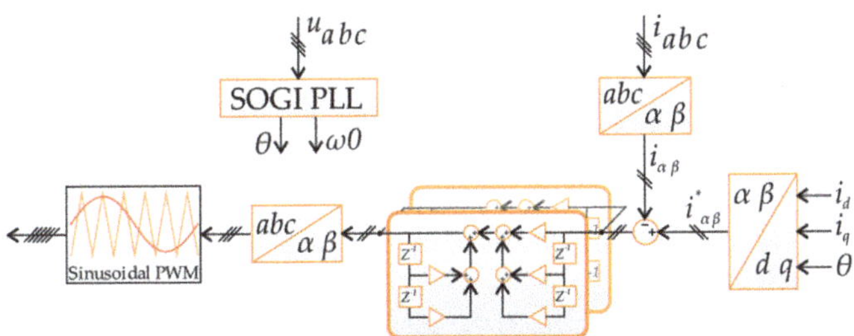

Figure 8. Implemented control structure. SOGI-PLL, second order generalized integrator phase-locked loop.

This synchronization method has the advantage of selectively rejecting all frequencies, except for a certain bandwidth adjusted by the gain k regardless of the resonance frequency w_n, without any lag.

Proposed Methodology

The proposed methodology is based on the grey wolf optimization (GWO) algorithm. It is a meta-heuristic algorithm proposed by [41] that mimics the social leadership and hunting behavior of grey wolves (agents). Its social leadership has a rigid and hierarchical structure, divided into four types of agents: the dominant agent is called alpha (α); second in the hierarchy is the beta agent (β); immediately below is the delta agent (δ); and finally, the lowest position in the social hierarchy is occupied by the omega agents (ω). Hunting behavior is divided into four stages: encircling prey, hunting, attacking prey, and searching for prey. The dynamics and harmony of these four stages establish the balance between the mechanisms of diversification and intensification and are coordinated through two control parameters, described by Equations (18) and (19):

$$a = 2 - 2\frac{t}{Max_{iter}} \tag{18}$$

$$A_d = 2ar - a \tag{19}$$

where r is a random number between $[0, 1]$, Max_{iter} is the number of maximum allowed iterations, and t is the current iteration. The searching for prey stage, which occurs when $|A_d| > 1$, forces agents to diverge from the best solution found so far (x_α), favouring the search for new solutions in unexplored regions (diversification mechanism). The remaining stages favour the intensification mechanism, forcing the construction of solutions in promising regions already explored. For a multidimensional search space, the new position of each agent is expressed by Equations (20)–(22).

$$D_{\alpha,d} = \left|C_d x_{\alpha,d}(t) - x_{n_a,d}(t)\right|, D_{\beta,d} = \left|C_d x_{\beta,d}(t) - x_{n_a,d}(t)\right|, D_{\delta,d} = \left|C_d x_{\delta,d}(t) - x_{n_a,d}(t)\right| \tag{20}$$

$$x_{1,d} = x_{\alpha,d}(t) - A_d D_{\alpha,d}, x_{2,d} = x_{\beta,d}(t) - A_d D_{\beta,d}, x_{3,d} = x_{\delta,d}(t) - A_d D_{\delta,d} \tag{21}$$

$$x_{n_a,d}(t+1) = \frac{x_{1,d} + x_{2,d} + x_{3,d}}{3} \tag{22}$$

where d represents the dimension of the search-space, n_a is the agent number of the population n_p, x_α is the position of the alpha agent, x_β is the position of the beta agent, x_δ is the position of the delta agent, and C_d is a random number within $[0, 2]$.

Figure 9 presents the flowchart of the proposed method, where, firstly, all variables and all parameters referring to GWO are initialized, such as the dimension of the problem, the control parameters, the lower (*lb*) and upper (*ub*) bounds, the number of agents of the population, and the iterations limit. The parameters for each configuration were confined within the boundaries indicated in Table 1. These boundaries were chosen in accordance with the literature to include a comprehensive set of solutions.

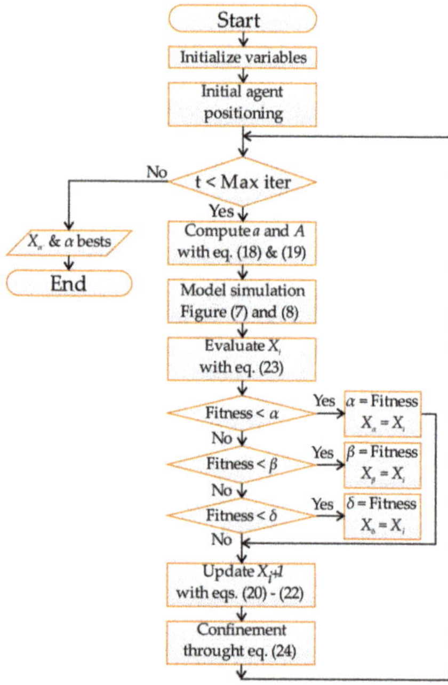

Figure 9. Fluxogram of the proposed method.

Table 1. Optimization boundaries for each output filter topology.

Parameters	Topology 1		Topology 2		Topology 3		Topology 4	
	Lower Bound	Upper Bound	Lower Bound	Upper Bound	Lower Bound	Upper Bound	Lower Bound	Upper Bound
L_1 (mH)	0.01	4	0.01	4	0.01	4	0.01	4
L_2 (mH)	0.01	4	0.01	4	0.01	4	0.01	4
L_d (mH)					0.01	4	0.01	4
C (μF)	1×10^{-5}	10	1×10^{-5}	10	1×10^{-5}	10	1×10^{-5}	10
C_d (μF)			1×10^{-5}	2.5			1×10^{-5}	0.5
R_d (Ω)	0.5	200	0.5	200	0.5	200	0.5	200
k_p	100	600	100	600	100	600	100	600
k_r	1×10^3	25×10^4	1×10^3	25×10^4	1×10^3	25×10^4	1×10^3	25×10^4

In the optimization process, 30 agents were used. After the initialization of the variables and parameters, a random initial positioning of the agents was performed within the search space.

The performance of each agent was evaluated through a fitness function (f_{obj}), where the problem was formulated as the minimization of Equation (23), subject to the restrictions shown in Table 1.

$$f_{obj} = \overbrace{\frac{\sqrt{\frac{1}{n}\sum_{i=1}^{n}\left(i_{\alpha,\beta} - i_{\alpha,\beta}^{*}\right)^2}}{max\left(i_{\alpha,\beta}^{*}\right)}}^{first\ term} + \overbrace{\frac{\sqrt{\sum_{h=2}^{\infty} i_h}}{i_1}}^{second\ term} + \overbrace{\frac{\sqrt{\sum_{i=1}^{n} P_{loss}}}{0.1P_{nom}}}^{third\ term} \tag{23}$$

The fitness function tries to achieve a complete symbiosis and harmony between three fundamental terms to maximize the efficiency of the system. The first term of Equation (23) quantifies the error between the current reference and the current measured in the stationary frame, through normalized root-mean-square error (NRMSE), allowing the determination of the optimum parameters of the PR current controller. The remaining terms of Equation (23) search a balance between antagonistic and fundamental terms for the correct design of the output filter, as they quantify the damping effect and power loss. Thus, the second term quantifies the total harmonic distortion (THD) present in the output current, allowing to measure the distortion of the current waveform. Finally, the third term quantifies the power loss in the damping resistor.

After the agents' evaluation, the three best in minimizing the f_{obj} and their respective position (x_α, x_β and x_δ) are determined and used to calculate the movement of the remaining population, as described above by Equations (20)–(22). To prevent agents from traveling outside the search space, during the successive iterations, the random positioning strategy was implemented. In this strategy, if any of the limits (lower or upper bounds) are exceeded, the movement of the agent is modified, ensuring that the new positioning is within the search space. This procedure is expressed by Equation (24).

$$x_{n_a,d}(t+1) = lb_d + (ub_d - lb_d)r \tag{24}$$

The recursive process ends as soon as the stopping criterion is reached. In particular, the stopping criterion used consists of the maximum number of permitted iterations ($Max_{iter} = 200$).

4. Simulation and Results

To validate the performance of the proposed method in determining the optimal parameters of the PR controllers and the optimal design of the output filter, two case studies were performed. The first case study was carried out with an SPD topology commonly found in the literature and the second case study with three different CPD topologies.

The system represented in Figures 7 and 8 was developed in Matlab/Simulink® simulation environment and the computing tasks were implemented on a computer with an Intel® Xeon® processor E5-1620 @ 3.60 GHz CPU, 8 GB RAM, and with Windows 10 Professional 64-bit operating system.

4.1. First Case Study—SPD Topology

The results obtained in the first case study (topology 1), using the proposed method, are presented in Table 2. The optimization process reached a fitness value of 0.3210, where almost 90% of the fitness value is related to NRMSE, which quantifies the error of the PR current controllers in both axes of the stationary frame. Both PR controllers present a similar performance. However, the PR controller in the α-frame has a slightly higher performance.

Table 2. Obtained parameters for topology 1. SPD, series passive damping.

Parameters	L_1 (mH)	L_2 (mH)	C (μF)	R_d (Ω)	k_r	k_p	f_{obj}
Topology SPD	4	0.723	0.868	4.63	86,245.556	180.198	0.3210

The second term of the fitness function that quantifies the THD value reached a value of 0.0195, which satisfies the constraints established by the IEEE-519 and IEEE-1547 standards. Finally, the third term of the fitness function quantifies the power loss in the damping resistor, and obtained a value of 0.0013, which corresponds to a root mean square (RMS) value of 1.5 W.

Figure 10a illustrates the frequency response of the output filter, optimized by the proposed method. It is possible to observe an attenuation of 60 dB/dec for high frequencies and an attenuation of 20 dB/dec for frequencies between 10 Hz and the resonance frequency (6.85 KHz). A slight amplification around the resonance frequency is visible, resulting from the trade-off between the damping effect and the power loss. In addition, the output filter stability is ensured with a gain margin of 38.11 dB and a phase margin of 113.57 degrees. Figure 10b shows the fast Fourier transform (FFT) of the output current, where a low switching noise around the switching frequency and below the resonance frequency can be noticed.

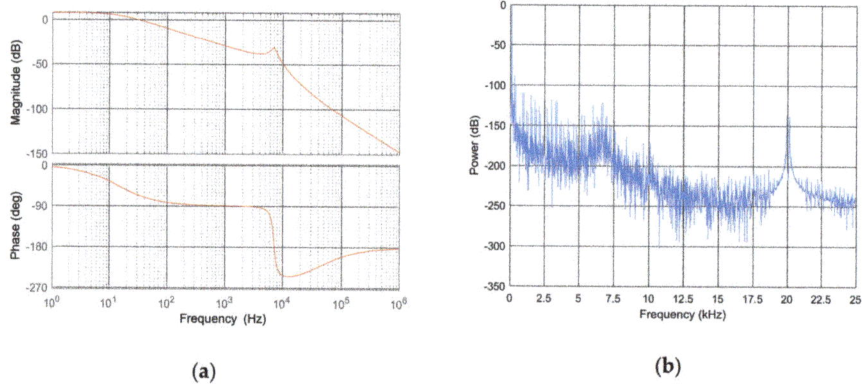

(a)   (b)

Figure 10. Output filter with topology 1, optimized with the proposed method. (**a**) Bode diagram; (**b**) fast Fourier transform.

To validate and compare the proposed methodology in the SPD topology, an analytical method was implemented for the design of the output filter, proposed by [42]. For this, an attenuation factor *ka* of 20% and a maximum allowed ripple on output current of 10% were considered. As a result, a value of 1.4 mH was obtained for L_1 inductor, 0.2 mH for L_2 inductor, 18.37 µF for capacitor C, and 0.3510 Ω for the damping resistance. Figure 11a shows the frequency response of the calculated filter, where a resonance frequency of 8.2 KHz, an attenuation for high frequencies of 60 dB/dec, a gain margin of 26.71 dB, and a phase margin of 89.99 degrees can be observed.

Additionally, for comparison purposes, in terms of FFT and THD, the same experimental procedure with the same operating conditions was performed. In Figure 11b, it is possible to observe the power spectrum of the output current obtained through the analytic method. When compared with the proposed method in terms of FFT and THD, it presents a similar noise around the switching frequency. However, it is possible to notice a lower attenuation in the harmonic components of the low frequency current. Another indicator of the signal harmonic distortion is the THD value, which in this case is 7.08%.

Figure 12 presents the values of the fitness function terms obtained by both methods. The proposed method achieves a 27% reduction in the fitness value. The most significant reduction occurred in the value of THD and in the current tracking error in the β-frame. In addition, the proposed method presents a lower power loss in the damping resistor, which is contradictory to the expected results. This is because the proposed method addresses both optimization problems (PR controllers and output filter design) as a single process, allowing a symbiosis and a better frequency response of the system.

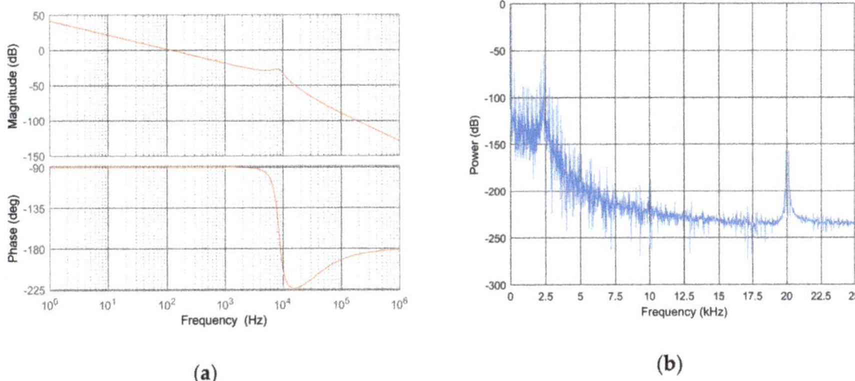

Figure 11. Output filter with topology 1, calculated through the conventional analytic method. (**a**) Bode diagram; (**b**) fast Fourier transform.

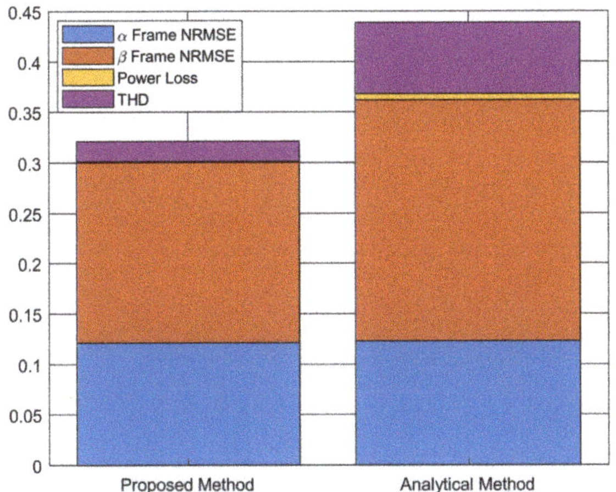

Figure 12. Fitness values obtained from the proposed method (left bar) and from the analytical method (right bar). NRMSE, normalized root-mean-square error; THD, total harmonic distortion.

4.2. Second Case Study—CPD Topologies

The results obtained for the CPD topologies using the proposed method are summarized in Table 3. The values of the fitness function show that all topologies obtained a similar performance. However, topology 2 achieved the best performance with a fitness value of 0.317; followed by topology 4 with a fitness value of 0.322; and, finally, topology 3 with a fitness value of 0.323. The NRMSE value obtained by the controllers in the α-frame is similar in all CPD topologies. Despite this, concerning the controllers in the β-frame, it is topology 4 that presents a lower NRMSE value. Regarding the total harmonic distortion, all topologies satisfy the constraints established by the IEEE-519 and IEEE-1547 standards. However, topology 2 stands out with a THD of 2.45%, followed by topology 4 with a THD of 2.73%, and lastly topology 3 with a value of 2.81%. Regarding the power loss in the damping resistor, the best configuration was topology 2 with an RMS value of 0.5 W, followed by topology 3 with an RMS value of 4.33 W, and finally topology 4 with an RMS value of 8.89 W.

Table 3. Obtained parameters for filter topologies 2, 3, and 4.

Parameters	L_1 (mH)	L_2 (mH)	L_d (mH)	C (µF)	C_d (µF)	R_d (Ω)	k_r	k_p	f_{obj}
Topology 2	3.7	0.135		4.71	1.68	2.42	78,945.05	173.13	0.317
Topology 3	3.8	0.729	0.41	6.08		82.3	93,109.50	199.24	0.323
Topology 4	3.78	2.44	2.91	3.86	2.72	172.73	129,451.34	287.27	0.322

Figure 13 illustrates the frequency responses of the output filters with the CPD topologies, optimized by the proposed method. It is possible to verify that both topologies have a similar behaviour in the attenuation of low frequency components. However, topology 4 presents a slight superiority in attenuating the components in that range. Regarding the attenuation of high frequency components, the behaviour of the three topologies is different. Topology 3 has the worst performance with an attenuation of 40 dB/dec. The remaining topologies (2 and 4) exhibit an attenuation of 60 dB/dec, although topology 4 presents a slightly superior performance in the attenuation of these components. Regarding the gain margin, topology 2 achieved a higher gain of 10.40 dB, followed by topology 4 with a gain of 6.05 dB, and lastly topology 3 with a gain of 5.06 dB. In terms of phase margin, all topologies present a similar value of 113 degrees.

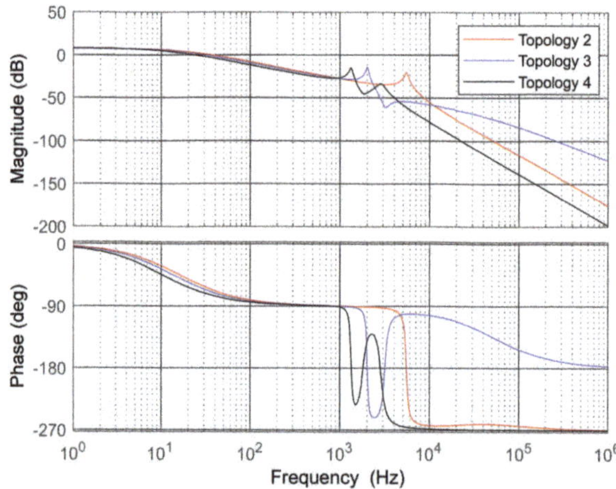

Figure 13. Bode diagram of the frequency response from filter with topologies 2, 3, and 4, obtained from the proposed method.

The fast Fourier transforms (FFTs) of the output currents of the three CPD topologies, optimized by the proposed method, are shown in Figure 14. Comparing the three FFTs, it is possible to verify that topology 4 exhibits the greatest attenuation for high frequency components, as shown in Figure 14c. The topology that offers the worst performance in this operating range is topology 3, as illustrated in Figure 14b. This is because topology 3 presents a 40 dB/dec attenuation for the high frequency components, as shown in Figure 13. In relation to THD, topology 2 has the best result with a value of 2.45%. The remaining topologies (3 and 4) present similar THD values with 2.81% and 2.73%, respectively.

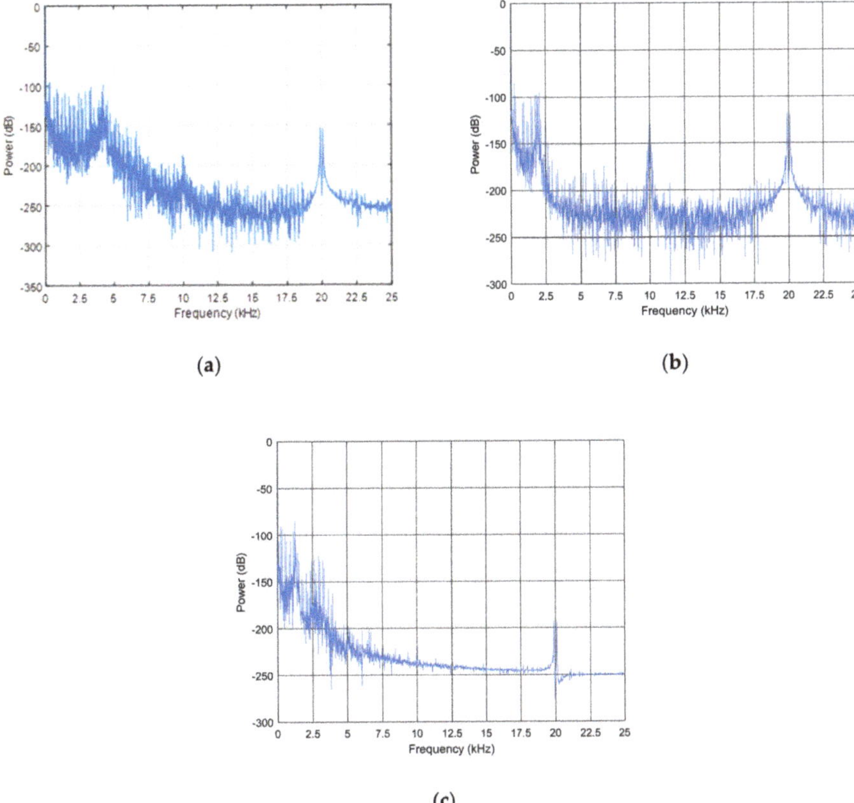

Figure 14. Fast Fourier transform of the frequency response for the three CPD filter topologies, obtained from the proposed method. (**a**) Topology 2; (**b**) topology 3; (**c**) topology 4.

Lastly, in Figure 15, it is possible to analyze the different terms of the fitness function for the three CPD topologies. All the three topologies reached similar fitness function values. However, topology 2 has the lowest value (0.317); followed by topology 4 with a value of 0.322; and, finally, topology 3 with a value of 0.323. Regarding the current tracking error, the NRMSE of the controllers in the α-frame is similar in all CPD topologies. However, the β-frame topology 4 exhibits a slightly lower NRMSE. The most expressive differences between each topologies are related to the power loss and the THD value. Once again, topology 2 stands out, reaching the lowest values in both power loss and THD.

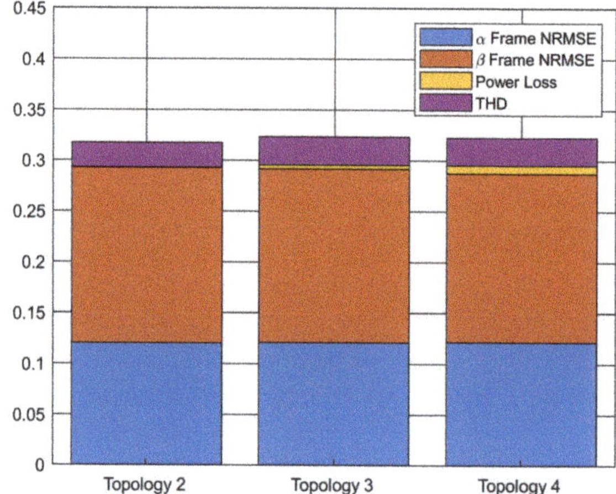

Figure 15. Fitness values obtained from the proposed method with topology 2 (left bar), topology 3 (centre bar), and topology 4 (right bar).

5. Conclusions

This paper proposed a new method to simultaneously determine the optimal control parameters of PR controllers and optimize the design of the output filter of a grid-tied three-phase inverter. The proposed method, based on the GWO algorithm, addresses both optimization problems as a single process, to achieve a complete symbiosis and a better system frequency response. It optimizes the parameters through an objective function that obtains the best trade-off between some fundamental terms: the harmonic attenuation rate, the power loss through the damping resistor, and the current tracking error. To validate the proposed methodology, two case studies were considered, with different output filter structures. The results obtained were analyzed according to three fundamental terms, in order to maximize the efficiency of the system. The first term quantifies the error between the current reference and the current measured in the stationary frame, through normalized root-mean-square deviation (NRMSE). The remaining terms quantify the total harmonic distortion and the power loss in the damping resistor. The different optimized topologies achieve an excellent performance, presenting a similar behavior. However, the results showed that topology 2 achieved a slightly superior performance, in terms of both THD and power loss in the damping resistor. In view of the above, it is possible to conclude the validity of the proposed methodology in both optimization problems.

Author Contributions: All authors contributed equally to this work. All authors have read and agreed to the published version of this manuscript.

Funding: This work is funded by FCT/MCTES through national funds and when applicable co-funded EU funds under the project UIDB/EEA/50008/2020.

Conflicts of Interest: The authors declare no conflict of interest.

References

1. Blaabjerg, F.; Yang, Y.; Yang, D.; Wang, X. Distributed Power-Generation Systems and Protection. *Proc. IEEE* **2017**, *105*, 1311–1331. [CrossRef]
2. Faria, J.; Pombo, J.; Calado, M.; Mariano, S. Power Management Control Strategy Based on Artificial Neural Networks for Standalone PV Applications with a Hybrid Energy Storage System. *Energies* **2019**, *12*, 902. [CrossRef]
3. Zakzouk, N.E.; Abdelsalam, A.K.; Helal, A.A.; Williams, B.W. High Performance Single-Phase Single-Stage Grid-Tied PV Current Source Inverter Using Cascaded Harmonic Compensators. *Energies* **2020**, *13*, 380. [CrossRef]

4. Faria, J.; Pombo, J.; Calado, M.d.R.; Mariano, S. Current Control Optimization for Grid-Tied Inverters Using Cuckoo Search Algorithm. In Proceedings of the International Congress on Engineering University da Beira Interior—"Engineering for Evolution", Covilhã, Portugal, 27–29 November 2019.

5. Committee, D.; Power, I.; Society, E. *IEEE Std 519-2014 (Revision of IEEE Std 519-1992)*; IEEE: Piscataway, NJ, USA, 2014; Volume 2014. [CrossRef]

6. Dong, R.; Liu, S.; Liang, G. Research on Control Parameters for Voltage Source Inverter Output Controllers of Micro-Grids Based on the Fruit Fly Optimization Algorithm. *Appl. Sci.* **2019**, *9*, 1327. [CrossRef]

7. Husev, O.; Roncero-Clemente, C.; Makovenko, E.; Pimentel, S.P.; Vinnikov, D.; Martins, J. Optimization and Implementation of the Proportional-Resonant Controller for Grid-Connected Inverter with Significant Computation Delay. *IEEE Trans. Ind. Electron.* **2020**, *67*, 1201–1211. [CrossRef]

8. Singh, J.K.; Behera, R.K. Hysteresis Current Controllers for Grid Connected Inverter: Review and Experimental Implementation. In Proceedings of the IEEE International Conference on Power Electronics, Drives and Energy Systems, PEDES 2018, Chennai, India, 18–21 December 2018; Institute of Electrical and Electronics Engineers Inc.: Piscataway, NJ, USA, 2018. [CrossRef]

9. Pouresmaeil, E.; Akorede, M.F.; Montesinos-Miracle, D.; Gomis-Bellmunt, O.; Trujillo Caballero, J.C. Hysteresis Current Control Technique of VSI for Compensation of Grid-Connected Unbalanced Loads. *Electr. Eng.* **2014**, *96*, 27–35. [CrossRef]

10. Qazi, S.H.; Mustafa, M.W.; Ali, S. Review on Current Control Techniques of Grid Connected PWM-VSI Based Distributed Generation. *Trans. Electr. Eng. Electron. Commun.* **2019**, *17*, 152–168. [CrossRef]

11. Bielskis, E.; Baskys, A.; Valiulis, G. Controller for the Grid-Connected Microinverter Output Current Tracking. *Symmetry* **2020**, *12*, 112. [CrossRef]

12. Kojabadi, H.M.; Yu, B.; Gadoura, I.A.; Chang, L.; Ghribi, M. A Novel DSP-Based Current-Controlled PWM Strategy for Single Phase Grid Connected Inverters. *IEEE Trans. Power Electron.* **2006**, *21*, 985–993. [CrossRef]

13. Moghaddam, A.F.; Van Den Bossche, A. Investigation of a Delay Compensated Deadbeat Current Controller for Inverters by Z-Transform. *Electr. Eng.* **2018**, *100*, 2341–2349. [CrossRef]

14. Naderipour, A.; Asuhaimi, A.; Zin, M.; Hafiz, M.; Habibuddin, B.; Miveh, M.R.; Guerrero, J.M.; Asuhaimi Mohd Zin, A.; Bin Habibuddin, M.H.; Miveh, M.R.; et al. An Improved Synchronous Reference Frame Current Control Strategy for a Photovoltaic Grid-Connected Inverter under Unbalanced and Nonlinear Load Conditions. *PLoS ONE* **2017**, *12*. [CrossRef] [PubMed]

15. Meral, M.E.; Çelik, D. Comparison of SRF/PI- and STRF/PR-Based Power Controllers for Grid-Tied Distributed Generation Systems. *Electr. Eng.* **2018**, *100*, 633–643. [CrossRef]

16. Golestan, S.; Ebrahimzadeh, E.; Guerrero, J.M.; Vasquez, J.C. An Adaptive Resonant Regulator for Single-Phase Grid-Tied VSCs. *IEEE Trans. Power Electron.* **2018**, *33*, 1867–1873. [CrossRef]

17. Yazdani, A.; Iravani, R. *Voltage-Sourced Converters in Power Systems: Modeling, Control, and Applications*; IEEE Press: Piscataway, NJ, USA; John Wiley: Hoboken, NJ, USA, 2010.

18. Kalaam, R.N.; Muyeen, S.M.; Al-Durra, A.; Hasanien, H.M.; Al-Wahedi, K. Optimisation of Controller Parameters for Gridtied Photovoltaic System at Faulty Network Using Artificial Neural Network-Based Cuckoo Search Algorithm. *IET Renew. Power Gener.* **2017**, *11*, 1517–1526. [CrossRef]

19. Qais, M.H.; Hasanien, H.M.; Alghuwainem, S. A Grey Wolf Optimizer for Optimum Parameters of Multiple PI Controllers of a Grid-Connected PMSG Driven by Variable Speed Wind Turbine. *IEEE Access* **2018**, *6*, 44120–44128. [CrossRef]

20. Guha, D.; Roy, P.K.; Banerjee, S. Grasshopper Optimization Algorithm-Scaled Fractional-Order PI-D Controller Applied to Reduced-Order Model of Load Frequency Control System. *Int. J. Model. Simul.* **2019**, 1–26. [CrossRef]

21. Dash, P.; Saikia, L.C.; Sinha, N. Automatic Generation Control of Multi Area Thermal System Using Bat Algorithm Optimized PD-PID Cascade Controller. *Int. J. Electr. Power Energy Syst.* **2015**, *68*, 364–372. [CrossRef]

22. Qi, Z.; Shi, Q.; Zhang, H. Tuning of Digital PID Controllers Using Particle Swarm Optimization Algorithm for a CAN-Based DC Motor Subject to Stochastic Delays. *IEEE Trans. Ind. Electron.* **2019**, *67*, 5637–5646. [CrossRef]

23. Khan, I.A.; Alghamdi, A.S.; Jumani, T.A.; Alamgir, A.; Awan, A.B.; Khidrani, A. Salp Swarm Optimization Algorithm-Based Fractional Order PID Controller for Dynamic Response and Stability Enhancement of an Automatic Voltage Regulator System. *Electronics* **2019**, *8*, 1472. [CrossRef]

24. Patel, N.C.; Debnath, M.K. Whale Optimization Algorithm Tuned Fuzzy Integrated PI Controller for LFC Problem in Thermal-Hydro-Wind Interconnected System. In *Lecture Notes in Electrical Engineering*; Springer: Berlin, Germany, 2019; Volume 553, pp. 67–77. [CrossRef]

25. Huang, M.; Li, H.; Wu, W.; Blaabjerg, F. Observer-Based Sliding Mode Control to Improve Stability of Three-Phase LCL-Filtered Grid-Connected VSIs. *Energies* **2019**, *12*, 1421. [CrossRef]

26. Beres, R.N.; Wang, X.; Liserre, M.; Blaabjerg, F.; Bak, C.L. A Review of Passive Power Filters for Three-Phase Grid-Connected Voltage-Source Converters. *IEEE J. Emerg. Sel. Top. Power Electron.* **2016**, *4*, 54–69. [CrossRef]

27. Gomes, C.C.; Cupertino, A.F.; Pereira, H.A. Damping Techniques for Grid-Connected Voltage Source Converters Based on LCL Filter: An Overview. *Renew. Sustain. Energy Rev.* **2018**, *81*, 116–135. [CrossRef]

28. Huang, M.; Blaabjerg, F.; Loh, P.C. The Overview of Damping Methods for Three-Phase Grid-Tied Inverter with LLCL-Filter. In Proceedings of the 16th European Conference on Power Electronics and Applications, Lappeenranta, Finland, 26–28 August 2014. [CrossRef]

29. Peña-Alzola, R.; Blaabjerg, F. Design and Control of Voltage Source Converters with LCL-Filters. In *Control of Power Electronic Converters and Systems*; Elsevier: Amsterdam, The Netherlands, 2018; pp. 207–242. [CrossRef]

30. Jalili, K.; Bernet, S. Design of LCL Filters of Active-Front-End Two-Level Voltage-Source Converters. *IEEE Trans. Ind. Electron.* **2009**, *56*, 1674–1689. [CrossRef]

31. Channegowda, P.; John, V. Filter Optimization for Grid Interactive Voltage Source Inverters. *IEEE Trans. Ind. Electron.* **2010**, *57*, 4106–4114. [CrossRef]

32. Kim, Y.-J.; Kim, H. Optimal Design of LCL Filter in Grid-Connected Inverters. *IET Power Electron.* **2019**, *12*, 1774–1782. [CrossRef]

33. Zhang, D.; Dutta, R. Application of Partial Direct-Pole-Placement and Differential Evolution Algorithm to Optimize Controller and LCL Filter Design for Grid-Tied Inverter. In Proceedings of the Australasian Universities Power Engineering Conference, AUPEC 2014, Perth, WA, Australia, 28 September–1 October 2014; pp. 1–6. [CrossRef]

34. Jianfeng, L.; Meiyu, L.; Guangzheng, Y.; Rusi, C. Parameter Optimization Design Method of LCL Filter Based on Harmonic Stability of VSC System. In Proceedings of the 4th International Conference on Intelligent Green Building and Smart Grid, IGBSG 2019, Yi-chang, China, 6–9 September 2019; pp. 226–231. [CrossRef]

35. Chatterjee, A.; Mohanty, K.B. Current Control Strategies for Single Phase Grid Integrated Inverters for Photovoltaic Applications—A Review. *Renew. Sustain. Energy Rev.* **2018**, *92*, 554–569. [CrossRef]

36. Liang, W.; Liu, Y.; Ge, B.; Wang, X. DC-Link Voltage Balance Control Strategy Based on Multidimensional Modulation Technique for Quasi-Z-Source Cascaded Multilevel Inverter Photovoltaic Power System. *IEEE Trans. Ind. Inform.* **2018**, *14*, 4905–4915. [CrossRef]

37. Zou, Z.-X.; Rosso, R.; Liserre, M. Modeling of the Phase Detector of a Synchronous-Reference-Frame Phase-Locked Loop Based on Second-Order Approximation. *IEEE J. Emerg. Sel. Top. Power Electron.* **2019**. [CrossRef]

38. Teodorescu, R.; Liserre, M.; Rodriguez, P. *Grid Converters for Photovoltaic and Wind Power Systems*; John Wiley & Sons, Ltd.: Hoboken, NJ, USA, 2011. [CrossRef]

39. Teodorescu, R.; Blaabjerg, F.; Liserre, M.; Loh, P.C. Proportional-Resonant Controllers and Filters for Grid-Connected Voltage-Source Converters. *IEE Proc. Electr. Power Appl.* **2006**, *153*, 750. [CrossRef]

40. Wu, W.; He, Y.; Tang, T.; Blaabjerg, F. A New Design Method for the Passive Damped LCL and LLCL Filter-Based Single-Phase Grid-Tied Inverter. *IEEE Trans. Ind. Electron.* **2013**, *60*, 4339–4350. [CrossRef]

41. Mirjalili, S.M.; Mirjalili, S.M.; Lewis, A. Grey Wolf Optimizer. *Adv. Eng. Softw.* **2014**, *69*, 46–61. [CrossRef]

42. Reznik, A.; Simoes, M.G.; Al-Durra, A.; Muyeen, S.M. LCL Filter Design and Performance Analysis for Grid-Interconnected Systems. *IEEE Trans. Ind. Appl.* **2014**, *50*, 1225–1232. [CrossRef]

Article

A Novel Lagrangian Multiplier Update Algorithm for Short-Term Hydro-Thermal Coordination [†]

P. M. R. Bento [1], S. J. P. S. Mariano [1,*], M. R. A. Calado [1] and L. A. F. M. Ferreira [2]

[1] IT—Instituto de Telecomunicações, University of Beira Interior, 6201-001 Covilhã, Portugal; pedro.bento@lx.it.pt (P.M.R.B.); rc@ubi.pt (M.R.A.C.)

[2] Instituto Superior Técnico and INESC-ID, University of Lisbon, 1049-001 Lisbon, Portugal; lmf@ist.utl.pt

*   Correspondence: sm@ubi.pt; Tel.: +351-(12)-7532-9760

†   This paper is an extended version of our paper published in ICEUBI2019—International Congress on Engineering—Engineering for Evolution, Covilhã, Portugal, 27–29 November 2019; pp. 728–742.

Received: 14 September 2020; Accepted: 11 December 2020; Published: 15 December 2020

Abstract: The backbone of a conventional electrical power generation system relies on hydro-thermal coordination. Due to its intrinsic complex, large-scale and constrained nature, the feasibility of a direct approach is reduced. With this limitation in mind, decomposition methods, particularly Lagrangian relaxation, constitutes a consolidated choice to "simplify" the problem. Thus, translating a relaxed problem approach indirectly leads to solutions of the primal problem. In turn, the dual problem is solved iteratively, and Lagrange multipliers are updated between each iteration using subgradient methods. However, this class of methods presents a set of sensitive aspects that often require time-consuming tuning tasks or to rely on the dispatchers' own expertise and experience. Hence, to tackle these shortcomings, a novel Lagrangian multiplier update adaptative algorithm is proposed, with the aim of automatically adjust the step-size used to update Lagrange multipliers, therefore avoiding the need to pre-select a set of parameters. A results comparison is made against two traditionally employed step-size update heuristics, using a real hydrothermal scenario derived from the Portuguese power system. The proposed adaptive algorithm managed to obtain improved performances in terms of the dual problem, thereby reducing the duality gap with the optimal primal problem.

Keywords: hydro-thermal coordination; Lagrangian relaxation; Lagrangian dual problem; Lagrange multipliers; subgradient methods; step-size update algorithm

---

1. Introduction

The objective of short-term hydro-thermal coordination is to optimize electricity generation [1], meaning to find an optimal generation dispatch, or close to ideal, for all the thermal and hydro units available in a system. This ensures the total operation cost is minimized within horizons ranging from one day to one week (168 h), taking into account the entire system and its individual constraints [2–5] and with a planning period (discrete time-step), typically set from hour to hour [5]. In other words, this crucial process is responsible for scheduling the start-up and shutdown of thermal units (binary level decisions), in coordination with hydro plants, to ensure the continuity of electricity supply with appropriate levels of spinning reserve, while minimizing the operating costs [6]. This scheduling constitutes a unit commitment (UC) problem, where the dispatch policy of the thermal units is made in such a way that the total cost (operating cost, starting cost and shut down cost) is minimal over a pre-defined time-horizon. In addition, a series of operational constraints needs to be fulfilled, thus reducing the freedom of choice to turn a thermal unit on or off. In this regard, we are primarily speaking about the load balance constraint, i.e., ensuring that our electric energy demand is satisfied, yet further constraints include spinning reserve constraints, minimum connected time, minimum time off, generation capacity limits, group restrictions, water restrictions, etc. [7].

Therefore, we can understand why short-term hydro-thermal coordination, in a framework where hydro and thermal power plants are the backbone of conventional power systems (consolidated power generation technologies), is such an important subject for power producers. Moreover, is one of the most complex problems to solve in power system engineering [1,8] due to its inherent large-scale and non-linear and combinatorial nature. This explains why, over the last decades, it has been the subject of intense research in the fields of operation and optimization of electric power systems [1,2,9,10].

Thus, a broad spectrum of methods has tried to solve short-term hydro-thermal coordination, and they can be generally divided into two categories: mathematical methods and deterministic methods [5,7]. In the realm of conventional approaches we can highlight Benders Decomposition [11], Lagrangian Relaxation [1,2,12], Improved LR [13], Dynamic Programming (PD), Nonlinear Programming [14], Dynamic Non-Linear Programming [15], Augmented Lagrangian [16], mixed integer linear programming [17–19], logarithmic size mixed-integer linear programming [20] and nonlinear programming [21], among several others. However, even with this wide array of classical optimization methods a perfectly tailored solution is hard to find, and in general terms the complexity of short-term hydro-thermal coordination has a negative effect on the computing efficiency [19]. Besides, problems of different nature arise with the use of classical approaches, which may impact the performance, mainly scheduling problems [22], slower convergence, premature convergence, computational cost, problems to deal with the nonlinearity and non-convexity, the need to perform problem simplifications, etc. In addition, classical deterministic methods typically rely upon single path search methods, which may help in terms of convergence speed but can be tricky in the presence of non-smooth surfaces [23].

A consolidated trend has been the growing application of evolutionary/heuristic methods and methods of artificial intelligence, in addition to new deterministic heuristics. Hence, we can mention neural networks [24], Cuckoo Search [25], Differential Evolution [26,27], Grey Wolf Optimizer [28], Improved Bacterial Foraging Algorithm [29] and a hybrid approach combining Artificial Bee Colony and the Bat algorithm [22], among others. For example, in [30] by using two different case studies, with and without pumping-storage capability and considering different constraints, the authors tested the effectiveness of different Accelerated Particle Swarm Optimization variants. In another instance, an Improved harmony search algorithm was employed on a non-linear, non-convex, short-term hydrothermal scheduling [31], among many others. However, in general these machine learning and population-based methods require a significant computational effort to solve the problem for an hourly discretized weeklong time-horizon, i.e., for large-scale problems (with a high number of dimensions) its effectiveness drops significantly. In addition, they can frequently end up finding only suboptimal solutions [4,21]. Besides, these metaheuristic methods often rely on a population search to find an optimal solution, turning them into large-scale problems (many dimensions and numbers of search agents), and occasionally several runs are required to find an optimal solution, as premature stagnation or slow convergence may occur.

Due to the presence of multiple sets of constraints, decomposition techniques appear as a natural option to solve this problem [1,10]. Consequently, Lagrangian Relaxation (LR) is one of these preferred decomposition techniques to tackle the short-term hydro-thermal coordination problem [6,12,32,33]. The fundamentals behind LR are to use Lagrange multipliers to relax system constraints such as load demand and reserve requirements. The primal problem is then converted into a two-level structure (subproblems). Given a set of multipliers, all subproblems are resolved at the low level, one for each unit, and the multipliers are updated at the high level. Multipliers are obtained by solving the dual problem, and the feasibility of solving the primal problem is usually obtained based on the dual solution [2,3], i.e., the primal problem is a byproduct of the dual problem solution. Finally, the solution is translated in dispatch generation decisions to meet the demand.

To update Lagrange multipliers, a common approach is to apply subgradient methods, where the step-size update procedure represents a sensible decision and often depends on ad-hoc testing.

This easy-to-implement approach, however, has some limitations, particularly the computational inefficiency and tendency for oscillating solutions. Two other common alternatives are the Bundle method, which in essence is another subgradient method based on a bundle of subgradients from previous iterations, and the Cutting Plane method, which adds new constraints to reduce the size of the feasible region. For example, in [34,35] evolutionary programming with a Gaussian mutation is used to update the multipliers, and in both, parameters are chosen considering convergence criteria. Following the vast existing literature on this subject, and to avoid sensible user dependence concerning the parameter choice, an adaptive algorithm is proposed in this work for an enhanced use of the LR technique for short-term hydro-thermal coordination. Additionally, it is important to refer that this paper is an extended version of work published in [36]; we expand on the formulation and the framework of the short-term hydro-thermal coordination problem, add a second case study, refine the results analysis and improve the figures.

Hence, we can summarize the main contributions of this work as follows:

1. Identify a weakness in the classical update mechanism of the step-size used in the subgradient method.
2. Propose an adaptative Lagrangian multiplier update algorithm that, as its name suggests, dynamically updates the step-size value and subsequently the Lagrange multipliers, so that the dual function converges towards its optimum in a pre-arranged number of iterations.
3. The LR technique is then used with the proposed adaptive update algorithm to solve a real large-scale, short-term hydro-thermal coordination problem, using data from the largest Portuguese electric utility company in two different scenarios.
4. For validation purposes, the algorithm is tested against two traditional step-size update heuristics with different initial parameter values.
5. Finally, the results comparison in both case studies revealed a sizeable advantage in terms of dual problem solution (error reduction) in favor of the proposed adaptative algorithm, therefore reducing the duality gap with the primal problem. Thus, a better allocation of the complex and vast hydro and thermal resources is obtained.

The rest of this paper is organized as follows. Section 2 provides a brief description of the primal problem. Section 3 explains the Lagrangian dual problem. Section 4 introduces subgradient methods and the motivation for the proposed algorithm, introduced in Section 5. Results and discussion for the two case studies are provided in Section 6, and finally, Section 7 presents the conclusion of this work.

2. The Primal Problem

The hydro-thermal coordination problem is a non-linear, large-scale, non-convex and combinatorial problem by nature. It can be understood as the task of establishing a map of feasible operations for each generation unit available in an electrical power system at the lowest cost for a predefined time horizon, in order to satisfy the expected load demand and a set of other system restraints. Typically, the time horizon considered is from one to seven days, and the discrete time-step (in which decisions are made) is a one-hour period. This problem is treated as deterministic, and whenever it is necessary to include stochastic quantities such as load diagram and reservoir inflows, their expected values are used.

In this manner, a primal problem ($\mathcal{P}$) is non-convex and non-linear and can be mathematically formulated as shown in Equations (1)–(6). The total operating cost for all resources (units) and over the entire considered period, K, is defined in Equation (1) and is the problem's objective function, i.e., evaluates the performance of each admissible solution. The cost function, $C_{ik}\left(x_{i,k-1}, p_{ik}, u_{ik}\right)$, is a measure that evaluates the decision made in each state, since there is an operating cost associated with the state transition (from $x_{i,k-1}$ to x_{ik}), which delivers the power p_{ik}, determined by the control decision u_{ik}, for each unit i at time k. The following three Equations (2)–(4) represent the set of global constraints. Firstly, Equation (2) translates the (global) demand–supply balance restraint, where D_k

is the required load demand that needs to be served by the power output of each resource i in hour k, p_{ik}. Moreover, for simplicity purposes transmission losses were not considered. In turn, Inequality (3) represents all the hourly system capacity requirement constraints, i.e., constraints like the spinning and operating reserve requirements. R_{mi} translates the capacity contribution function associated with resource i for the system capacity requirement of type m, while R_{mk}^{req} is the mth-type system capacity requirement in hour k [5].

Additionally, Inequality (4) represents all the cumulative constraints, such as the limitation on emission by a group of units over the scheduling time horizon or the amount of consumed fuel, where $\mathcal{H}_n$ stands for the set of thermal units on the nth cumulative constraint, H_{ni} is the function which describes a contribution of thermal unit i–nth cumulative constraint, H_n^{req} is the upper bound on nth cumulative constraint and N is the set of cumulative constraints [37].

In turn, Equations (5) and (6) represent the set of local constraints, the state Equation (5) of each resource i at a time k. This equation allows us to obtain the state of each resource x_{ik} and its contribution p_{ik} to satisfy demand, whatever the decision u_{ik}. Last of all, in (6) the domain of admissible values for the control variables, as well as for the initial and final state, are defined for each individual resource i.

$$\mathcal{P} \quad \underset{u}{Min} \quad \sum_{k=1}^{K}\sum_{i=1}^{I} C_{ik}\left(x_{i,k-1}, p_{ik}, u_{ik}\right) \tag{1}$$

Subject to

$$\sum_{i=1}^{I} p_{ik} = D_k \quad i=1,\ldots,I \wedge k=1,\ldots,K \tag{2}$$

$$\sum_{i=1}^{I} R_{mi}\left(x_{ik}, p_{ik}\right) \geq R_{mk}^{req} \quad m=1,\ldots,M \tag{3}$$

$$\sum_{k=1}^{K}\sum_{i\in\mathcal{H}_n} H_{ni}\left(x_{ik}, p_{ik}, u_{ik}\right) \geq H_n^{req} \quad n=1,\ldots,N \tag{4}$$

and wherein

$$\left(x_{ik}, p_{ik}\right) = A_{ik}\left(x_{i,k-1}, u_{ik}\right) \quad i=1,\ldots,I \wedge k=1,\ldots,K \tag{5}$$

$$u_{ik} \in \mathcal{U}_{ik} \quad x_{i0} \in X_i^0 \quad x_{ik} \in X_i^K$$
$$i=1,\ldots,I \quad \wedge \quad k=1,\ldots,K \tag{6}$$

Although the objective function is a separable function in resources and hours, this problem, by its formulation and due to collective constraints, does not allow this separability, providing extreme complexity to the minimization problem. In other words, the optimum value cannot be found by the sum of the various suboptimal (separately) results from each resource. Thus, we are facing a problem of unrestrainable dimension, for which a direct approach is not viable.

The primal problem defined in this study approaches the short-term hydro-thermal coordination considering the generation resources available to the electric utilities company, to match the system-wide load demand over a weekly time-period, while fulfilling a set of other global and local constraints.

3. Lagrangian Dual Problem

As discussed, the primal problem is difficult to solve using conventional nonlinear optimization techniques. A preferable path is to decompose the problem constraints and transfer them to the objective function, i.e., to solve the dual problem, rather than solving the primal problem directly. We know beforehand that the optimal solution of the relaxed problem is a lower bound (good estimate) of the optimal solution of the initial problem [2,10,38].

This is achieved by relaxing the constraints, i.e., weakening of the problem ($\mathcal{P}$), that in practical terms means open the possibility to breach the imposed constraints. However, relaxed restrictions are not completely ignored since its violations are linearly penalized in the Lagrange function (an added cost associated with violating each constraint).

We can write the Lagrange function ($\mathcal{L}$) for problem ($\mathcal{P}$) by relaxing its global constraint, as expressed in Equation (7), where λ, μ and γ are the Lagrange multiplier vectors associated with the load-balance constraint, capacity constraints and cumulative constraints, respectively. These Lagrange multipliers are expressed in units of cost per unit of the parameters of their associated constraint, which in the case of Equation (2) is given in \$/GW.

$$
\begin{aligned}
\mathcal{L}\big(x_{i,k-1}, p_{ik}, u_{ik}, \lambda, \mu, \gamma\big) = {} & \sum_{k=1}^{K} \sum_{i=1}^{I} C_{ik}\big(x_{i,k-1}, p_{ik}, u_{ik}\big) \\
& + \sum_{k=1}^{K} \lambda_k \bigg(D_k - \sum_{i=u}^{I} p_{ik} \bigg) \\
& + \sum_{m=1}^{M} \sum_{k=1}^{K} \mu_{mk}\bigg(R_{mk}^{req} - \sum_{i=1}^{I} R_{mi}(x_{ik}, p_{ik}) \bigg) \\
& + \sum_{n=1}^{N} \gamma_n \bigg(H_n^{req} - \sum_{k=1}^{K} \sum_{i \in \mathcal{H}_n} H_{ni}(x_{ik}, p_{ik}, u_{ik}) \bigg)
\end{aligned}
\tag{7}
$$

That is, to now solve the unit commitment problem, $\mathcal{L}$ is minimized, where $\underset{u}{Min}\ \mathcal{L}\big(x_{i,k-1}, p_{ik}, u_{ik}, \lambda, \mu, \gamma\big)$ is subject to local system constraints, i.e., Equations (5) and (6).

Subgradient of the Dual Function

The Lagrangian dual problem is obtained by forming ($\mathcal{L}$), and its solution provides the primal variables as functions of the Lagrange multipliers, which are labeled dual variables. Hence, the new problem is to maximize the objective function with respect to the multipliers under the derived constraints on the dual variables. This implies, by decomposition, that each resource becomes a single entity and is individually optimized, rather than a combined optimal resource allocation. Therefore, the dual function is defined in Equation (8), presenting concave and sub-differentiable traits (resulting in inferiorly limited variables).

$$
q(\lambda, \mu, \gamma) = \underset{u}{Min}\ \mathcal{L}\big(x_{i,k-1}, p_{ik}, u_{ik}, \lambda, \mu, \gamma\big)
\tag{8}
$$

Given that Lagrange's dual function is a concave function with simple bounds on the variables, a local optimum is also the function global optimum. Therefore, our task is to find the Lagrangian multipliers, λ, μ and γ, that maximize the dual function. Nonetheless, this does not mean that solving the dual function is a trivial task (far from it actually), since the function is not smooth and is not given by an easy-to-compute expression [5]. For this purpose, we resort to subgradient methods, which benefit from the fact the subgradients of q are easily derived system constraint deviations. Consequently, we can define the subgradient of the dual function g (9) for each hour k as follows:

$$g = \begin{bmatrix} \cdots \\ D_k - \sum_{i=1}^{I} p_{ik} \\ \cdots \\ \hline \\ \cdots \\ R_{mk}^{req} - \sum_{i=1}^{I} R_{mi}(x_{ik}, p_{ik}) \\ \cdots \\ \hline \\ \cdots \\ H_n^{req} - \sum_{k=1}^{K} \sum_{i \in \mathcal{H}_n} H_{ni}(x_{ik}, p_{ik}, u_{ik}) \\ \cdots \end{bmatrix} \tag{9}$$

Moreover, by the weak duality theorem for a single set of multipliers, the optimal value of the Lagrange dual problem $q(\lambda^*)$ and the optimal value of the primal minimization problem $p(\lambda^*)$ are related by $q(\lambda^*) \leq p(\lambda^*)$, and the difference between the values is called a duality gap. This implies that the dual problem offers a good indirect root to solve the primal one, since the gap in most practical cases is relatively small [6,12,39].

For all the reasons above, this approach to the problem is advantageous since it lessens the computational burden of the primal problem.

4. Subgradient Methods

As we saw earlier, obtaining the Lagrange dual function optimal value goes hand-in-hand with the Lagrange multiplies choice/update method, i.e., at the outset, this choice determines how close we are to the solution of the dual problem and, ultimately, how close we are from reaching the primal problem best solution. To perform this task several methods are described in the literature [5]; however, regarding our problem in particular, subgradient methods prevail as the most fitting solution by achieving higher accuracies. Further benefits include their simplicity as well as the computational ease with which the Lagrange dual function subgradient (solution deviation from the imposed constraints) is calculated.

These methods update the multipliers according to the subgradient direction and in a manner proportional to the violation of the corresponding constraints. Besides, a distinctive trademark of these methods concerns the step-size update heuristic, where again several approaches have been followed [5]. However, the downside of these conventional updating heuristics is that a long-winded trial-and-error procedure as well as a highly specialized operator are frequently required. The simplest and most common subgradient method formulation is given by

$$\lambda^{v+1} = \left[\lambda^v + s^v \frac{g^v}{\|g^v\|} \right]^+ \tag{10}$$

where g^v is the subgradient $g(p_{\lambda^v})$, s^v is a positive scalar that defines the step-size at the current iteration v and, lastly, $[.]^+$ represents the projection in the set of feasible values Λ. Nonetheless, there is no guarantee that after iteration $v + 1$, independently from the chosen step-size, the dual function value will actually improve (walk towards the optimal dual function value), meaning that in some occasions we will have

$$q\left(\left[\lambda^v + s^v \frac{g^v}{\|g^v\|} \right]^+ \right) < q(\lambda^v), \ \forall \ s > 0 \tag{11}$$

Though, if the step value is sufficiently small, the distance between the obtained point in the current iteration and the optimum value can always be reduced. The following proposition offers an estimate for the step-size domain (range):

Proposition 1. *If λ^v does not lead to the optimum value of the dual function, then for λ^*, which corresponds to the dual function optimum value, the inequality $\|\lambda^{v+1} - \lambda^*\| < \|\lambda^v - \lambda^*\|$ is valid for all step-sizes, $s^v \in \left]0, \frac{2(q(\lambda^*)-q(\lambda^v))}{\|g^v\|}\right[$. Therefore, this suggests a step-size equal to the middle value of the inequality range, i.e., $s^v = \frac{(q(\lambda^*)-q(\lambda^v))}{\|g^v\|}$.*

Since this requires knowledge of the dual function optimal value $q(\lambda^*)$, which is exactly the unknown we want to find, this approach is unviable in our case, and we resort to heuristics that determine the step-size. In this regard, a popular choice is decreasing step-size rule-based approaches, mainly due to its simplicity and effectiveness.

Considering a decrease in step-size, s^v, towards zero, meaning that $\lim_{v \to \infty} s^v = 0 \wedge s^v > 0$, while at the same time satisfying $\sum_{v=1}^{\infty} s^v = \infty$, the method can "travel" as far as possible (up to infinity) in order to converge to the optimal dual function value. Thus, under these assumptions, we can infer a 2nd proposition, from which we can conclude that it is possible, by appropriately updating the step-size, to reach the dual function maximum value [21].

Proposition 2. *For the sequence of all multiplier values $\{\lambda^v\}$ we have $\lim_{v \to \infty} Max\, q(\lambda^v) = q^*$. However, this analysis does not lead to a finite procedure, pointing to an initial value of the step, as well as a mechanism for decrementing it to zero. As such, for comparison purposes against the proposed new heuristic, the two most widely employed expressions are introduced in Equations (12) and (13), to update the step-size at each iteration v.*

$$s^v = \frac{a_1}{1 + v \times a_2} \tag{12}$$

$$s^v = \frac{a_1}{1 + v^{a_2}} \tag{13}$$

where a_1 and a_2 are control parameters of the heuristic process. Moreover, the chosen initial step is a highly sensitive matter, since small initial steps can prevent the method from reaching the desired optimum value in a reasonable number of iterations. Whereas, using a large initial step may cause the method to oscillate erratically in the early phase, leading to poor convergence. As a result, although the obtained value is stabilized, it could still be improved by running further iterations. This fact is more pronounced in Equation (12) rather than Equation (13), given that $a_2 > 1$ implies a rapid decrease in the step-size. Consequently, selecting the values to assign to parameters a_1 and a_2 is a difficult task, with direct influence on the obtained results. This could be facilitated if a good initial estimate for the dual variable vector λ^0 is available, that is, if $q(\lambda^0)$ is already close to the solution of the dual problem.

Therefore, we can conclude that it is an intrinsically lengthy (experimentation-based) heuristic process that is highly dependent on the user's experience. Precisely to mitigate this scenario, a new algorithm will be proposed next.

5. Proposed Adaptative Algorithm

When applying subgradient methods, the existence of good estimates for the multipliers and careful tune-up of the subgradient step-size are considered essential to improve the computational effectiveness of the method. Subsequently, motivated by the previously exposed shortcomings from the classical subgradient optimization approaches, an adaptative heuristic is proposed in order to automatically update the Lagrange multipliers, thereby removing the need to rely on a user's past experiences or time-consuming trial-and-error tasks. This means that the step-size, s^v, is automatically determined (avoiding lengthy trial-and-error procedures) by the adaptative algorithm when solving the dual problem with a subgradient method. The different stages of the algorithm that lead to a dual problem solution are illustrated in Figure 1; subsequently, the rationale behind them is detailed below:

(1) define the initial step-size, $s^0 = 1$, and choose an initial value for the dual variable vector, λ_0; then compute the initial dual function and subgradient values, $q^0(\lambda^0)$ and $g^0(p_{\lambda^0})$, respectively;

(2) update Lagrange multipliers according to Equation (10);

(3) determine the new step-size as follows:

if $q^v(\lambda^v) > q^{v-1}(\lambda^{v-1})$ then

$$\alpha \in v_\delta^+(1)$$

else

$$\alpha \in v_\delta^-(1)$$

end, where,

$$v_\delta^+(1) = \{\alpha_1 : 1 < \alpha < 1 + \delta\}$$

$$v_\delta^-(1) = \{\alpha_2 : 1 - \delta < \alpha < 1\}$$

and the step-size is given by

$$s^v = \alpha s^{v-1}$$

(4) compute the current (iteration) dual function and subgradient values, $q^v(\lambda^v)$ and $g^v(p_{\lambda^v})$;

(5) if the termination criterion is met:

a. terminate the algorithm;

else

b. proceed to the next iteration, $v = v + 1$;

c. return to (2);

end

Regarding the (above) adaptative algorithm, the following clarifications are made:

In (1) the initial dual variable vector positioning only impacts the convergence speed of the subgradient method; thereby, it can be considered arbitrary. On the contrary, for the step-size update expressed by Equations (12) and (13), this initial positioning benefits heavily from a nearby optimum value (derived from past experiences or other heuristics) in order to guarantee the method's performance, thereby translating an important advantage of the proposed strategy.

In turn, stage (3) depicts the original step-size update mechanism, where the rationale behind it is to dynamically update the step based on the dual function value, i.e., if this value improves then the step should be augmented; in contrast, if this value does not improve then the step should be diminished. Moreover, to prevent a large step-size increasing the distance between the new point and the optimum value, this step-size should be increased smoothly; this fact is less sensitive when reducing step-size. Additionally, it was found that the ideal domains for variables α_1 and α_2 are $[1.01, 1.05] \Rightarrow \delta = [0.01, 0.05]$ and $[0.83, 0.95] \Rightarrow \delta = [0.05, 0.17]$, respectively.

Lastly, the stop criterion mentioned in (4) is traditionally run a specific number of iterations, which was also the case in this work.

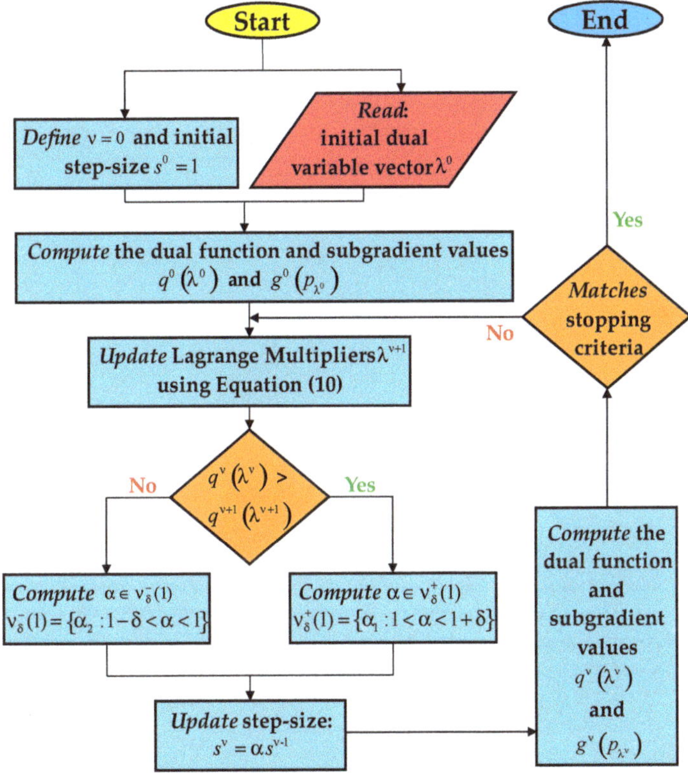

Figure 1. Flowchart of the proposed adaptative algorithm.

6. Numerical Results

The behavior of the subgradient method is analyzed in this section. The step value is updated according to the adaptive algorithm, proposed in Section 5, and then benchmarked against a classical approach. The step-size is updated using Equations (12) and (13) and, consequently, the Lagrange multipliers. The software used in this work was written in Fortran using the development environment (IDE) Microsoft Visual Studio ®.

As previously mentioned, the unit commitment (primal problem) corresponding to the solution of the Lagrange dual problem does not always lead to a feasible solution. As such, the average subgradient norm, $\|g(p_\lambda)\|/K$, is defined as a quantitative metric of how a solution is accurate in terms of the primal problem. This means the lower the value, the closer we will be to a good solution, and typically a value on the order of 0.5% of the peak load typically means that a good solution to the primal problem was found.

The data employed in this work concern the real short-term hydro-thermal coordination problem that the main Portuguese electric utility companies face. Data include all generation parameters and auxiliary variables, i.e., a large-scale study comprising six thermal power plants and 26 hydro power plants (amounting to over 80 individual generation units), which serve the majority of the Portuguese electric power demand. Two different case studies will be considered, diverging over the selected weekly periods, size and characteristics of the system, as well as the economic strategies behind the cost curves. Additionally, the specified parameters will be kept fixed for both case studies.

6.1. Case Study I

For this case study, unit costs are the sole result from the associated generation costs, with no other parallel costs. Consequently, both the evolution of the dual function $q(\lambda)$ value as well the evolution of the average subgradient norm $\|g(p_\lambda)\|/K$ will be evaluated over the course of iterations. Figures 2a,b and 3 show the evolution of the dual function value (left axis) and its step value (right axis). Figures 4a,b and 5 show the evolution of the average subgradient norm. Figures 2a and 4a illustrate the behavior of the subgradient method using a classical step-size update, given by Equation (12). In the same fashion, Figures 2b and 4b illustrate the behavior when using Equation (13). Lastly, the new adaptive algorithm results are shown in Figures 3 and 5.

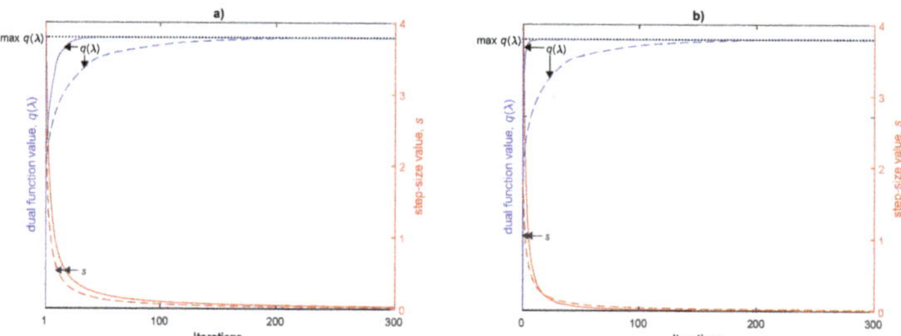

Figure 2. Evolution of the dual function value (blue plots), $q(\lambda)$, and its step value (red plots), using the heuristics expressed by Equations (12) and (13) for (a) and (b), respectively. The following parameter values are imposed: (**a**) solid line, $a_1 = 20$, $a_2 = 2$; dashed line, $a_1 = 10$, $a_2 = 1.5$; (**b**) solid line, $a_1 = 20$, $a_2 = 1.5$; dashed line, $a_1 = 5.5$, $a_2 = 1.05$.

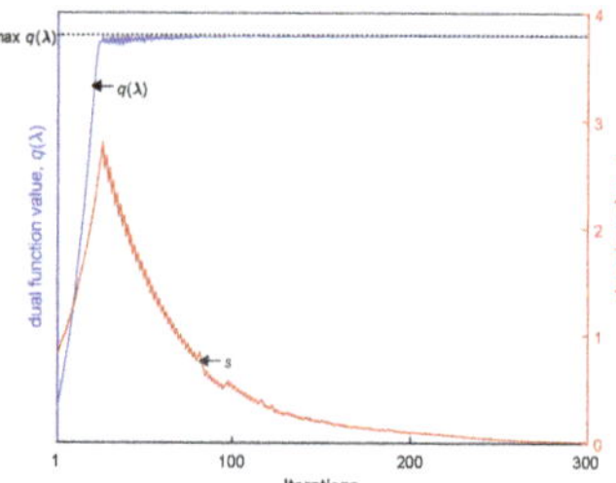

Figure 3. Evolution of the dual function value (blue plot), $q(\lambda)$, and its step value (red plot), using the adaptative algorithm, with the following parameter values: $\alpha_1 = 1.05$, $\alpha_2 = 1.10$.

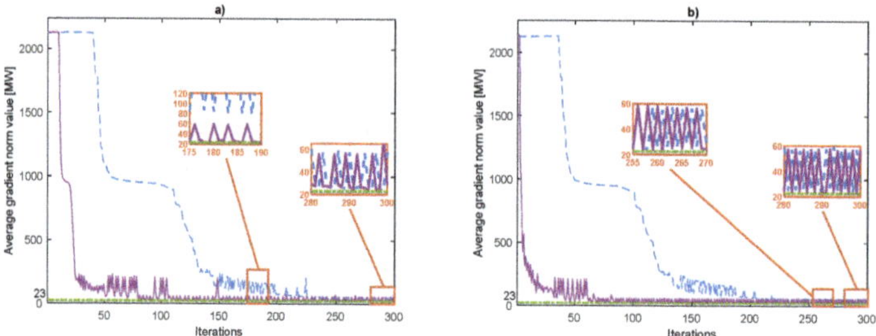

Figure 4. Evolution of the average subgradient norm, $\|g(p_\lambda)\|/K$, corresponding to the values of the dual function represented in Figure 2 (classical step-size equations). (**a**) purple solid line, $a_1 = 20$, $a_2 = 2$; blue dashed line, $a_1 = 10$, $a_2 = 1.5$; (**b**) purple solid line, $a_1 = 20$, $a_2 = 1.5$; blue dashed line, $a_1 = 5.5$, $a_2 = 1.05$. The green dot-dashed line highlights the achieved minimum gradient norm value.

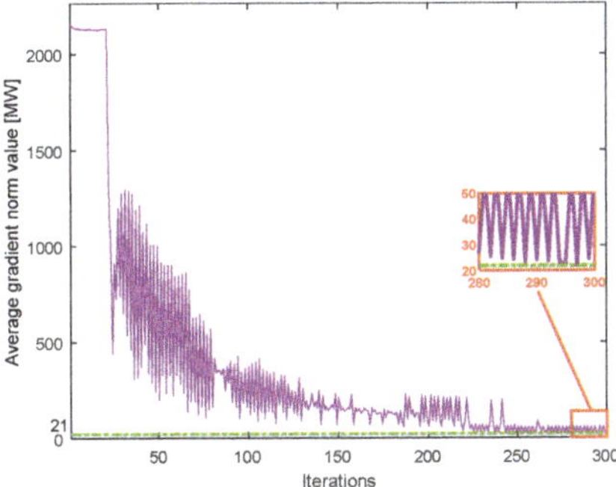

Figure 5. Evolution of the average subgradient norm (purple solid line), $\|g(p_\lambda)\|/K$, corresponding to the values of the dual function represented in Figure 3 (proposed adaptative algorithm). The green dot-dashed line highlights the achieved minimum gradient norm value.

Regarding Figure 2a,b we can observe the following: (i) achieving convergence in more or less iterations is heavily dependent on the choice of the different parameters; (ii) using a smaller initial step-size increased the number of iterations needed to achieve convergence, $\max q(\lambda)$ (dashed line); (iii) the use of a slightly larger initial step leads to some oscillation, still, without compromising convergence, represented by the solid lines; (iv) the step-size evolution is strictly decreasing, and the rate of descent depends on the considered parameters.

With respect to the adaptive algorithm, the evolution of the dual function increases as the value of the step increases, as shown in Figure 3, until a value is reached in the vicinity of the maximum dual function value. From this point onwards, the step value decreases towards zero, but then again it slightly increases whenever the dual function value does not improve compared to the previous iteration. This dynamically adjusted (based on the dual function current value) step-size clearly contrasts with the monotone evolution that occurs with the traditional step update formulation.

We can verify that, in all scenarios, the dual function maximum value was reached, and differences reside in the number of iterations necessary to achieve convergence (which was relatively similar). Nevertheless, the proposed algorithm shows a more robust approach when applying the subgradient method since it did not require educated guesses to converge on an acceptable number of iterations.

Concerning the obtained minimum average subgradient norm, Figures 4 and 5 are presented, where the secondary axis (magnified) provides a greater resolution regarding the convergence of each error curve to its recorded minimum value. Additionally, to summarize the respective minimum values achieved by the different step-size update mechanisms with different initial parameters, Table 1 is also presented.

Table 1. Minimum average subgradient norm, $\|g(p_\lambda)\|/K$, and its corresponding iteration, achieved by the different step-size update mechanisms for case study I.

Min $\frac{\|g(p_\lambda)\|}{K}$ [MW]	Iteration	Step-Size Update Mechanism
22.97	186	Equation (12) with $a_1 = 20$ and $a_2 = 2$
24.59	299	Equation (12) with $a_1 = 10$ and $a_2 = 1.5$
22.49	261	Equation (13) with $a_1 = 20$ and $a_2 = 1.5$
23.74	298	Equation (13) with $a_1 = 5.5$ and $a_2 = 1.05$
20.95	297	proposed adaptative algorithm

As we can see in Figure 4a,b, both processes led to similar final results, with a (best) value close to 23 MW by employing both classical step-size update equations. Nevertheless, the first combination in Table 1 ensured the fastest convergence (186 iterations). As for the proposed Lagrangian multiplier update algorithm, we noticed an improved (smaller) error of ~21 MW, which in this relatively curtailed error scenario represents an improvement above 8%. These values represent approximately 0.53% of the peak load, which, as mentioned, usually leads to a good solution to the primal problem. Besides, note that once the dual function maximum value or its proximities are reached, the average subgradient norm has not yet reached its minimum value, and it continues to oscillate up and down over the next iterations. This can be explained since small variations in the multipliers can cause large variations in the solutions in terms of the primal problem. Thus, emphasizing the fact that even when reaching the maximum value of the dual function, we may not end up with the best solution in terms of the primal problem.

Moreover, this average subgradient norm oscillation is further accentuated in the adaptative algorithm since, in contrast to the previous two step-size update expressions, it does not have a strictly decreasing behavior. This behavior can lead to a lower convergence speed and, therefore, constitutes a limitation of the proposed adaptive algorithm. Notwithstanding, it is worth noting that a slow convergence is preferred over a premature convergence.

Lastly, regarding the solution in terms of the primal problem (Figure 6), corresponding to the solution of the dual problem for the (achieved) lowest average subgradient norm value. The same was obtained using the adaptive algorithm, since is easy to understand from previous figures that all primary solutions would be similar, so their presentation is redundant. The algorithm used in solving the primal problem based on Lagrangian relaxation, as we saw earlier, does not lead to an optimal solution. The obtained primal solution reveals the existence of Lagrangian duality. That is, we can say that good results were obtained since the generation profile (solid green line) was almost coincident with the desired load demand profile (dashed red line), but it did not match it completely. After solving the dual problem, several methods have been used to look for feasibility [5]. However, if we succeed when solving the dual problem, then we can also get, in terms of the primal problem, a good solution. In fact, in some cases it is enough to carry out an economic dispatch of thermal units to obtain a (close) feasible strategy, which is exactly what happens in the presented case study, where the difference between the maximum generation capacity (dotted orange line) and the allocated thermal unit generation (dashed magenta line) is sufficient to compensate for the mismatch (deviation) between the load profile and the obtained generation profile.

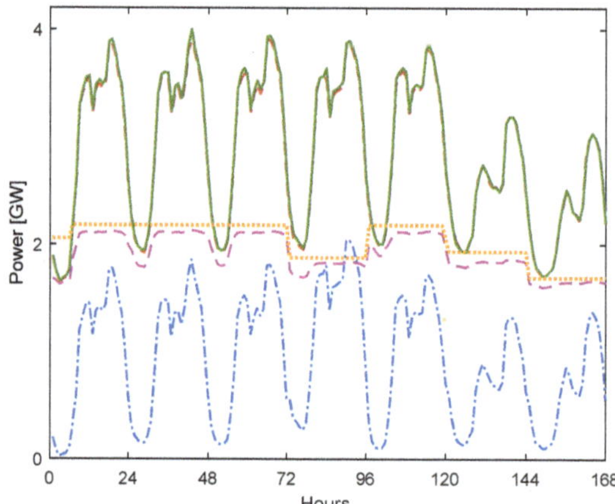

Figure 6. Solution in terms of the primal problem (case study I). In the upper portion the solid green and dashed red lines are almost coincident: the obtained generation profile and the load demand, respectively. Dotted orange line: maximum generation capacity of the affected thermal units. Dashed magenta line: thermal units generation profile. Dash-dot blue line: hydro units generation profile.

6.2. Case Study II

A second case study is considered with the purpose of further validating the proposed adaptive algorithm. For this scenario, the units' cost curves result is not the sole result of the generation costs but is also from additional economic strategies. Furthermore, the peak demand power (delivered to the grid) was 45% higher than in case study I. Thus, it constitutes a case with greater dimensions, with extra hydro and thermal plants considered. However, the parameters required for Equations (12) and (13) and the adaptive algorithm are the same as those specified in case study I, so we can compare the performance between these different methods of updating the step-size value.

In the same manner as in case study I, we started by analyzing the evolution of the dual function value, $q(\lambda)$, and the correspondent step-size values. From looking at Figure 7a,b we can see that the maximum value of the dual function, using both classical step-size update equations, had not been reached. Indeed, when compared to the value obtained using the adaptive algorithm (Figure 8), this value falls short by roughly 0.7% for the parameters illustrated by the solid line, whereas for the ones represented by dashed lines an evident lack of convergence can be spotted.

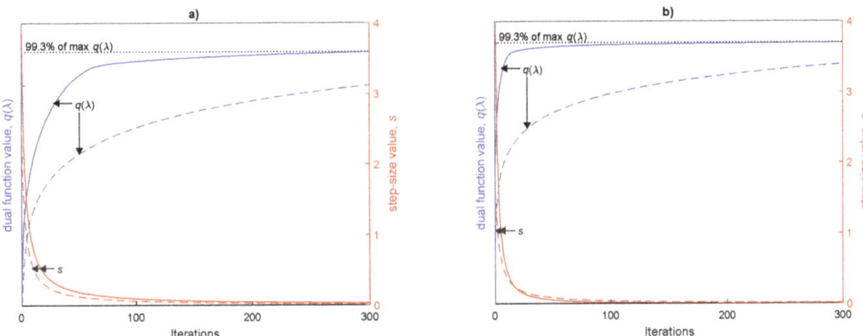

Figure 7. Evolution of the dual function value (blue plots), $q(\lambda)$, and its step value (red plots), using heuristic expressed by Equations (12) and (13) for (a) and (b), respectively. The following parameter values are imposed: (**a**) solid line, $a_1 = 20$, $a_2 = 2$; dashed line, $a_1 = 10$, $a_2 = 1.5$; (**b**) solid line, $a_1 = 20$, $a_2 = 1.5$; dashed line, $a_1 = 5.5$, $a_2 = 1.05$, respectively.

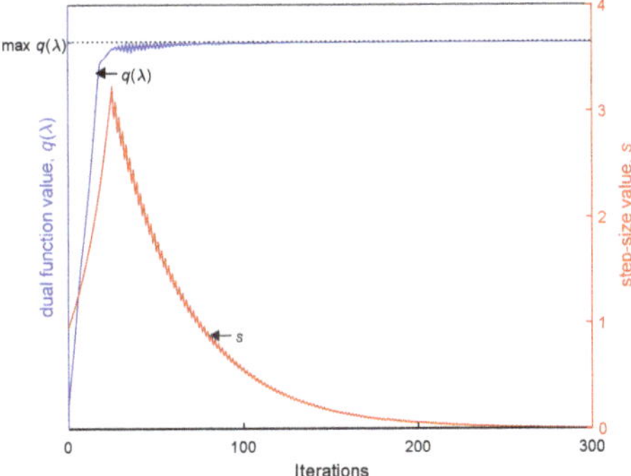

Figure 8. Evolution of the dual function value (blue plot), $q(\lambda)$, and its step value (red plot), using the adaptative algorithm, with the following parameter values: $\alpha_1 = 1.05$, $\alpha_2 = 1.10$.

In turn, the adaptative algorithm updates the step-size value dynamically and adapts to the current dual function value. That is, as mentioned above, if the dual function value improves then the step should be increased; on the contrary, if this value does not improve then the step should be decreased. Thus, in Figure 8 we can see that the step value increased until the dual function value approached its maximum value; thereafter, the step value was modified accordingly, which enabled convergence with the max$q(\lambda)$ in both case studies, and ultimately led to good primal solutions.

These results anticipate a higher minimum value of the average subgradient norm for the classical update expressions, which inevitably compromises the solution in terms of the primal problem. This effect should be particularly pronounced for the parameters represented by the dashed plots in Figure 7a,b. Therefore, a clear contrast to the results from the first case is established, where the maximum value of the dual function is relatively similar for the different mechanisms used to update the Lagrange multipliers (as shown in Table 1).

As expected, this preliminary assessment is fully backed by examining the evolution of the average subgradient norm, as shown below in Figure 9a,b, as well as in the summary Table 2, which presents the respective minimum values achieved by the different step-size update mechanisms with different initial parameters, represented by each individual error plot in Figures 9 and 10. The results reveal sizeable minimum average subgradient norm values, even for the scenarios where the dual function closed in on its maximum value (~99.3% of the max$q(\lambda)$), presenting values of 318 MW and 317 MW, respectively. These results differ from the ones achieved using the adaptive algorithm (Figure 10), where a much improved $\|g(p_\lambda)\|/K$ minimum value was recorded, 38 MW (~8.3 times smaller), i.e., roughly representing only 0.64% of the peak power demand versus >5% with the classic heuristics.

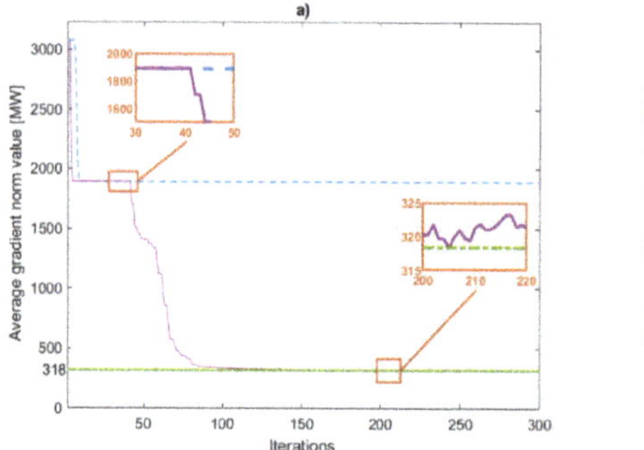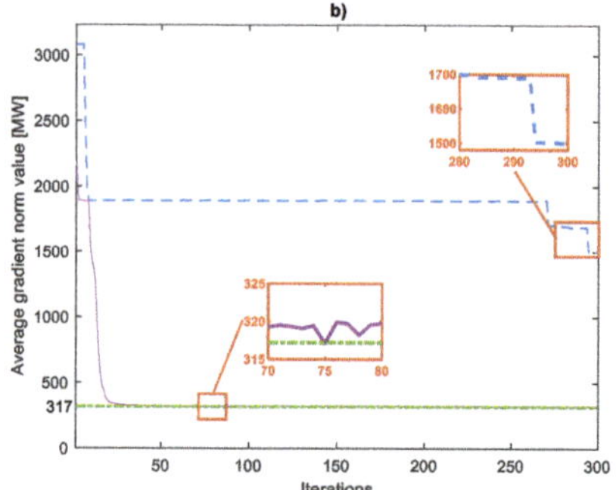

Figure 9. Evolution of the average subgradient norm, $\|g(p_\lambda)\|/K$, corresponding to the values of the dual function represented in Figure 7 (classical step-size equations). (**a**) Purple solid line, $a_1 = 20$, $a_2 = 2$; blue dashed line, $a_1 = 10$, $a_2 = 1.5$; (**b**) purple solid line, $a_1 = 20$, $a_2 = 2$; blue dashed line, $a_1 = 5.5$, $a_2 = 1.05$. The green dot-dashed line highlights the achieved minimum gradient norm value.

Table 2. Minimum average subgradient norm, $\|g(p_\lambda)\|/K$, and its corresponding iteration, achieved by the different step-size update mechanisms for case study II.

Min $\frac{\|g(p_\lambda)\|}{K}$ [MW]	Iteration	Step-Size Update Mechanism
318.33	205	Equation (12) with $a_1 = 20$ and $a_2 = 2$
1889.21	33	Equation (12) with $a_1 = 10$ and $a_2 = 1.5$
317.17	75	Equation (13) with $a_1 = 20$ and $a_2 = 1.5$
1499.30	299	Equation (13) with $a_1 = 5.5$ and $a_2 = 1.05$
38.37	278	proposed adaptative algorithm

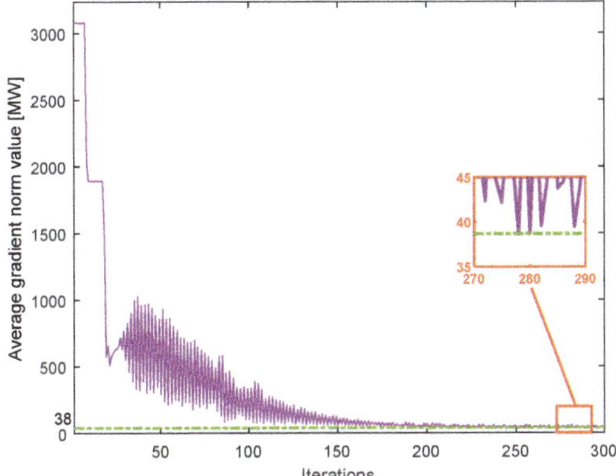

Figure 10. Evolution of the average subgradient norm (purple solid line), $\|g(p_\lambda)\|/K$, corresponding to the values of the dual function represented in Figure 8 (proposed adaptive algorithm). The green dot-dashed line highlights the achieved minimum gradient norm value.

Furthermore, we can also notice that some of the parameters led to a fast but also premature convergence when using traditional heuristics, contrasting with the slower convergence exhibited by the proposed adaptative algorithm due to a more refined step-size update. Thereby, the inferences highlighted for case study I are confirmed. Despite the significant improvement by several orders of magnitude compared to the more modest improvement in case study I, this improved error value is still slightly above the one obtained in the previous case study, which translates to added difficulty when considering additional dispatch strategies. These characteristics will ultimately impact the task of obtaining a feasible solution in terms of the primal problem. Nevertheless, the proposed adaptative algorithm was able to converge to a good solution, and, in comparison, it will result in a smaller duality gap. Besides, it reinforces the mentioned need to thoroughly adjust the control parameters or rely upon educated guesses in order to achieve a good result when using classic heuristics versus the proposed automated algorithm.

Moving on towards the primal problem solution, Figure 11 is presented, revealing a higher peak demand as well as a larger usage of the thermal capacity in direct comparison to its counterpart in case study I, Figure 6. Further, it can be noted that the hydro generation follows the load profile on a smaller scale, and the hydro-thermal generation profile obtained moves further away from the load profile. Thus, the economic dispatch of thermal units is not sufficient for the solution, in terms of the primal problem, to be feasible. In other words, contrary to the first case study, the difference between the maximum generation capacity (dotted orange plot) and the allocated thermal units' generation (dashed magenta plot), illustrated in Figure 11, is insufficient to address the mismatch between the load profile and the obtained generation profile. Additionally, we can see that around 1–3 h and 147–149 h the hydro generation registered a negative value, which is explained by a sporadic period of power consumption (pumping) during an off-peak occurrence.

For this reason, it makes sense to present, for this case study, the mismatch between the load profile and the obtained generation profile as illustrated in Figure 12, where we can observe the non-feasibility after the economic dispatch for a few hours. This behavior was more significant during the last considered day (between 144 and 168 h), which is a weekend day, reaching a maximum above 65 MW at time $k = 149$ h. This translates a mismatch around 1% of the peak demand, and it is justified by the limited availability of thermal generation (orange and magenta lines in Figure 11 almost overlapping during this period). Nonetheless, on average, the primal problem mismatch never exceeded 0.5% of the peak load demand.

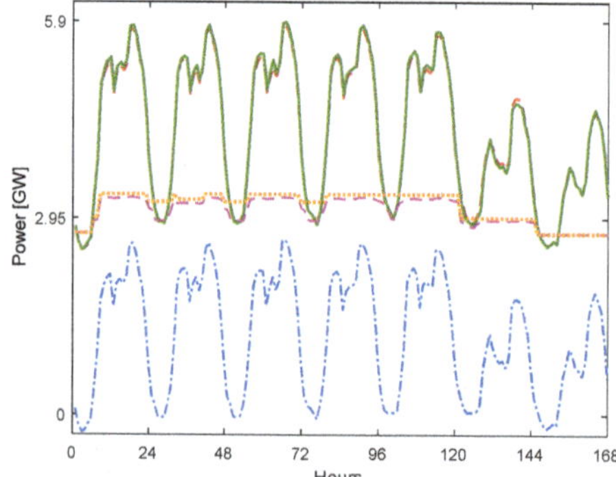

Figure 11. Solution in terms of the primal problem (case study II). In the upper portion are almost coincident solid green and dashed red lines: the obtained generation profile and the load demand, respectively. Dotted orange line: maximum generation capacity of the affected thermal units. Dashed magenta line: thermal units generation profile. Dash-dot blue line: hydro units generation profile.

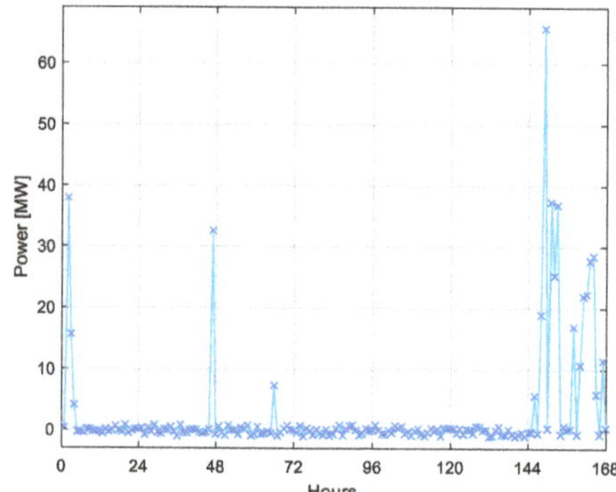

Figure 12. Primal problem solution mismatch: non-feasibility periods of the thermal units' economic dispatch.

7. Conclusions

In this paper, we explored the numerical performance of a novel Lagrangian multiplier update algorithm for short-term hydro-thermal coordination. The step-size update mechanism is a vital component for these methods, and classic approaches are heavily dependent upon a user's experience and fine-tuning procedures, i.e., selecting the appropriate parameters. The proposed algorithm had an important advantage of not requiring parameter choices based on experimentation, and it is subsequently compared against two classical update expressions. After a results assessment of both case studies, we could infer that the adaptive algorithm produced considerably improved dual problem solutions, seen through the percentage gains in terms of average subgradient norm and the respective ratio between the average subgradient norm and peak demand. This ultimately means an improved upper bound for the primal problem solution. Moreover, for most hours it led to feasible primal solutions, which in turn translates into more cost-effective dispatch decisions. The significance of the obtained results is magnified, especially when considering the differences in data prediction, such as the expected inflows.

These improvements proved the algorithm's ability to dynamically adapt the step value according to the dual function value. We can see that during the initial iterations, the step value is incremented until approaching the vicinities of the dual function maximum. From then onwards, the step evolves dynamically and adapts to the current dual function value, allowing convergence to the maximum dual function value and to the average subgradient norm, which translates to a feasible or a near-feasible primal solution. On the contrary, when using the classical update step-size update equations, it is necessary to rely on educated guesses or perform adjustments over the control parameters (as illustrated by case study II), through a trial-and-error process, in order to obtain a solution similar to the one achieved by the new adaptative algorithm.

Finally, we see that for a high-dimensional optimization problem, the computational burden is not exaggerated and not dependent upon initial guesses, especially considering that a weekly unit commitment is performed.

Author Contributions: Conceptualization, S.J.P.S.M. and L.A.F.M.F.; Methodology, S.J.P.S.M. and L.A.F.M.F.; Investigation, P.M.R.B., M.R.A.C., L.A.F.M.F. and S.J.P.S.M.; Software, S.J.P.S.M. and L.A.F.M.F.; Formal Analysis, P.M.R.B. and S.J.P.S.M.; Visualization, P.M.R.B., S.J.P.S.M. and M.R.A.C.; Supervision, S.J.P.S.M.; Writing—Original Draft Preparation, P.M.R.B., S.J.P.S.M. and M.R.A.C.; Writing—Review & Editing, M.R.A.C., P.M.R.B. and S.J.P.S.M. All authors have read and agreed to the published version of the manuscript.

Funding: This work is funded by FCT/MCTES through national funds and, when applicable, co-funded EU funds under the project UIDB/50008/2020.

Acknowledgments: P.M.R. Bento gives his special thanks to the Fundação para a Ciência e a Tecnologia (FCT), Portugal, for the Ph.D. grant (SFRH/BD/140371/2018).

Conflicts of Interest: The authors declare no conflict of interest.

Nomenclature

k	discrete time-step (an hour)
i	ith generation resource
K	total number of hours
I	total number of resources
C_{ik}	cost function associated with resource allocation i at time k
x_{ik}	resource state i at time k
p_{ik}	power output by resource i at time k
u_{ik}	control (decision) variable for resource i at time k
D_k	load demand at time k
m	mth-type system capacity requirement
R_{mi}	capacity contribution function associated with resource i for system capacity requirement of type m
R_{mk}^{req}	mth-type system capacity requirement in hour k
M	total number of capacity requirement constraints
n	nth cumulative constraint
$\mathcal{H}_n$	set of all resources constrained by nth cumulative constraint
H_{ni}	function of the contribution of resource i to the nth cumulative constraint
H_n^{req}	lower bound on the nth cumulative constraint
N	number of cumulative constraints
A_{ik}	state function associated with each resource i at time k
$\mathcal{U}_{ik}$	control variables (decision) universe for resource i at time k
X_i^0	resource i initial state
X_i^K	resource i final state
$\mathcal{L}$	Lagrange function
λ	Lagrange multiplier vector associated with the load-balance constraint
μ	Lagrange multiplier vector associated with the capacity constraints
γ	Lagrange multiplier vector associated with the cumulative constraints
$q(\lambda, \mu, \gamma)$	Lagrange dual function
g	subgradient of the dual function
v	current iteration of the subgradient method
s	step-size of the subgradient method
a_1	control parameter of the classic step update heuristic
a_2	control parameter of the classic step update heuristic
α_1	control parameter of the novel adaptative step update heuristic
α_2	control parameter of the novel adaptative step update heuristic
δ	indirect upper and lower bound parameter of the control parameters α_1 and α_2, respectively
$\|g(p_\lambda)\|/K$	average subgradient norm

References

1. Takigawa, F.Y.K.; Finardi, E.C.; da Silva, E.L. A decomposition strategy to solve the short-term hydrothermal scheduling based on Lagrangian relaxation. In Proceedings of the 2010 IEEE/PES Transmission and Distribution Conference and Exposition: Latin America (T&D-LA), Sao Paulo, Brazil, 8–10 November 2010; pp. 681–688.
2. Ye, H.; Zhai, Q.; Ge, Y.; Wu, H. A revised subgradient method for solving the dual problem of hydrothermal scheduling. In Proceedings of the 2011 Asia-Pacific Power and Energy Engineering Conference, Wuhan, China, 25–28 March 2011; pp. 1–6.
3. Zhai, Q.; Wu, H. Several notes on Lagrangian relaxation for unit commitment. In Proceedings of the 29th Chinese Control Conference, Beijing, China, 29–31 July 2010; pp. 1848–1853.
4. Nguyen, T.T.; Vo, D.N. Multi-Objective short-term fixed head hydrothermal scheduling using augmented lagrange hopfield network. *J. Electr. Eng. Technol.* **2014**, *9*, 1882–1890. [CrossRef]
5. Ferreira, L.A.F.M.; Andersson, T.; Imparato, C.F.; Miller, T.E.E.; Pang, C.K.K.; Svoboda, A.; Vojdani, A.F.; Irnparato, C.F.; Miller, T.E.E.; Pang, C.K.K. Short-Term resource scheduling in multi-area hydrothermal power systems. *Int. J. Electr. Power Energy Syst.* **1989**, *11*, 200–212. [CrossRef]
6. Jiménez Redondo, N.; Conejo, A.J. Short-term hydro-thermal coordination by lagrangian relaxation: Solution of the dual problem. *IEEE Trans. Power Syst.* **1999**, *14*, 89–95. [CrossRef]
7. Singhal, P.K. Generation scheduling methodology for thermal units using lagrangian Relaxation. In Proceedings of the 2011 Nirma University International Conference on Engineering, Gujarat, India, 8–10 December 2011; pp. 1–6.
8. Başaran Filik, Ü.; Kurban, M. Solving unit commitment problem using modified subgradient method combined with simulated annealing algorithm. *Math. Probl. Eng.* **2010**, *2010*, 1–15. [CrossRef]
9. Marmolejo-Saucedo, J.A.; Rodríguez-Aguilar, R. A proposed method for design of test cases for economic analysis in power systems. *J. Appl. Res. Technol.* **2015**, *13*, 428–434. [CrossRef]
10. Diniz, A.L.; Sagastizabal, C.; Maceira, M.E.P. Assessment of Lagrangian relaxation with variable splitting for hydrothermal scheduling. In Proceedings of the 2007 IEEE Power Engineering Society General Meeting, Tampa, FL, USA, 24–28 June 2007; pp. 1–8.
11. Lopez-Salgado, C.J.; Ano, O.; Ojeda-Esteybar, D.M. Hydrothermal scheduling with variable head hydroelectric plants: Proposed strategies using benders decomposition and outer approximation. In Proceedings of the 2016 IEEE Power and Energy Conference at Illinois (PECI), Champaign, IL, USA, 19–20 February 2016; pp. 1–8.
12. Rodrigues, R.N.; da Silva, E.L.; Finardi, E.C.; Takigawa, F.Y.K. Solving the short-term scheduling problem of hydrothermal systems via lagrangian relaxation and augmented Lagrangian. *Math. Probl. Eng.* **2012**, *2012*, 1–18. [CrossRef]
13. Cho, P.; Tang, Z.; Gong, H.; Zhao, X. An improved Lagrangian relaxation algorithm for the robust generation self-scheduling problem. *Math. Probl. Eng.* **2018**, *2018*, 1–12. [CrossRef]
14. Kumar, R.; Garg, V.; Lal, B. A review paper on hydro-thermal scheduling. *Int. J. Emerg. Technol. Comput. Appl. Sci.* **2013**, *5*, 522–526.
15. Hoseynpour, O.; Mohammadi-Ivatloo, B.; Nazari-Heris, M.; Asadi, S. Application of dynamic non-linear programming technique to non-convex short-term hydrothermal scheduling problem. *Energies* **2017**, *10*, 1–17. [CrossRef]
16. Beltran, C.; Heredia, F.J. Short-Term hydrothermal coordination by augmented Lagrangean relaxation: A new multiplier updating. In Proceedings of the IX Congreso Latino-Iberoamericano de Investigación Operativa, Buenos Aires, Argentina, 31 August–4 September 1998; pp. 1–6.
17. Gil, E.; Araya, J. Short-Term hydrothermal generation scheduling using a parallelized stochastic mixed-integer linear programming algorithm. *Energy Procedia* **2016**, *87*, 77–84. [CrossRef]
18. Fu, B.; Ouyang, C.; Li, C.; Wang, J.; Gul, E. An improved mixed integer linear programming approach based on symmetry diminishing for unit commitment of hybrid power system. *Energies* **2019**, *12*, 833. [CrossRef]
19. Kounalakis, M.E.; Theodorou, P. A hydrothermal coordination model for electricity markets: Theory and practice in the case of the Greek electricity market regulatory framework. *Sustain. Energy Technol. Assess.* **2019**, *34*, 77–86. [CrossRef]
20. Jian, J.; Pan, S.; Yang, L. Solution for short-term hydrothermal scheduling with a logarithmic size mixed-integer linear programming formulation. *Energy* **2019**, *171*, 770–784. [CrossRef]

21. Mariano, S.J.P.S.; Catalão, J.P.S.; Mendes, V.M.F.; Ferreira, L.A.F.M. Optimising power generation efficiency for head-sensitive cascaded reservoirs in a competitive electricity market. *Int. J. Electr. Power Energy Syst.* **2008**, *30*, 125–133. [CrossRef]

22. Ghosh, S.; Kaur, M.; Bhullar, S.; Karar, V. Hybrid ABC-BAT for solving short-term hydrothermal scheduling problems. *Energies* **2019**, *12*, 551. [CrossRef]

23. Farhat, I.A.; El-Hawary, M.E. Optimization methods applied for solving the short-term hydrothermal coordination problem. *Electr. Power Syst. Res.* **2009**, *79*, 1308–1320. [CrossRef]

24. Basu, M. Hopfield neural networks for optimal scheduling of fixed head hydrothermal power systems. *Electr. Power Syst. Res.* **2003**, *64*, 11–15. [CrossRef]

25. Nguyen, T.T.; Vo, D.N. Modified cuckoo search algorithm for short-term hydrothermal scheduling. *Int. J. Electr. Power Energy Syst.* **2015**, *65*, 271–281. [CrossRef]

26. Basu, M. Improved differential evolution for short-term hydrothermal scheduling. *Int. J. Electr. Power Energy Syst.* **2014**, *58*, 91–100. [CrossRef]

27. Zhang, H.; Zhou, J.; Zhang, Y.; Lu, Y.; Wang, Y. Culture belief based multi-objective hybrid differential evolutionary algorithm in short term hydrothermal scheduling. *Energy Convers. Manag.* **2013**, *65*, 173–184. [CrossRef]

28. Das, S.; Bhowmik, D.; Das, S. *A Modified Grey Wolf Optimizer Algorithm for Economic Scheduling of Hydrothermal Systems*; Springer: Cham, Switzerland, 2020; pp. 669–677.

29. Farhat, I.A.; El-Hawary, M.E. Short-Term hydro-thermal scheduling using an improved bacterial foraging algorithm. In Proceedings of the 2009 IEEE Electrical Power & Energy Conference (EPEC), Montreal, QC, Canada, 22–23 October 2009; pp. 1–5.

30. Fakhar, M.S.; Kashif, S.A.R.; Ain, N.U.; Hussain, H.Z.; Rasool, A.; Sajjad, I.A. Statistical performances evaluation of APSO and improved APSO for short term hydrothermal scheduling problem. *Appl. Sci.* **2019**, *9*, 2440. [CrossRef]

31. Nazari-Heris, M.; Babaei, A.F.; Mohammadi-Ivatloo, B.; Asadi, S. Improved harmony search algorithm for the solution of non-linear non-convex short-term hydrothermal scheduling. *Energy* **2018**, *151*, 226–237. [CrossRef]

32. Takigawa, F.Y.K.; Da Silva, E.L.; Finardi, E.C.; Rodrigues, R.N. Solving the hydrothermal scheduling problem considering network constraints. *Electr. Power Syst. Res.* **2012**, *88*, 89–97. [CrossRef]

33. Petcharaks, N.; Ongsakul, W. Hybrid enhanced Lagrangian relaxation and quadratic programming for hydrothermal scheduling. *Electr. Power Compon. Syst.* **2007**, *35*, 19–42. [CrossRef]

34. Padmini, S.; Jegatheesan, R.; Thayyil, D.F. A novel method for solving multi-objective hydrothermal unit commitment and sheduling for GENCO using hybrid LR-EP technique. *Procedia Comput. Sci.* **2015**, *57*, 258–268. [CrossRef]

35. Nayak, C.; Rajan, C.C.A. An evolutionary programming embedded Tabu search method for hydro-thermal scheduling with cooling banking constraints. *J. Eng. Technol. Res.* **2013**, *5*, 21–32. [CrossRef]

36. Bento, P.; Pina, F.; Mariano, S.; do Rosario Calado, M. Short-Term Hydro-Thermal Coordination by Lagrangian Relaxation: A New Algorithm for the Solution of the Dual Problem. Available online: https://knepublishing.com/index.php/KnE-Engineering/article/view/7093/12770 (accessed on 14 September 2020).

37. Catalão, J.P.S.; Mariano, S.J.P.S.; Mendes, V.M.F.; Ferreira, L.A.F.M. A practical approach for profit-based unit commitment with emission limitations. *Int. J. Electr. Power Energy Syst.* **2010**, *32*, 218–224. [CrossRef]

38. Ruiic, S. Optimal distance method for Lagrangian mulitpliers updating in short-term hydro-thermal coordination—Power systems. *IEEE Trans. Power Syst.* **1998**, *13*, 1439–1444.

39. Habibollahzadeh, H.; Brannlund, H.; Bubenko, J.; Sjelvgren, D.; Andersson, N. Optimal Short-Term Planning of Hydro-Thermal Power System—Part II: Solution Techniques. Available online: https://www.sciencedirect.com/science/article/pii/B9780408014687500527 (accessed on 14 September 2020).

Publisher's Note: MDPI stays neutral with regard to jurisdictional claims in published maps and institutional affiliations.

Article

New HVAC Sustainability Index—TWI (Total Water Impact)

Alexandre F. Santos [1,2,3,*], Pedro D. Gaspar [1,3,*] and Heraldo J. L. de Souza [2,*]

1 Department of Electromechanical Engineering, University of Beira Interior, 6201-001 Covilhã, Portugal
2 FAPRO—Faculdade Profissional, 80230-040 Curitiba, Brazil
3 C-MAST—Centre for Mechanical and Aerospace Science and Technologies, 6201-001 Covilhã, Portugal
* Correspondence: projetos.etp@gmail.com (A.F.S.); dinis@ubi.pt (P.D.G.); heraldosouza1@gmail.com (H.J.L.d.S.); Tel.: +55-413-332-7025 (A.F.S.); +351-275-329-759 (ext. 3759) (P.D.G.); +55-419-997-48928 (H.J.L.d.S.)

Received: 24 January 2020; Accepted: 27 March 2020; Published: 1 April 2020

Abstract: Sales of air conditioning are growing rapidly in buildings, more than tripling between 1990 and 2016. This energy use for air conditioning comes from a combination of rising temperatures, rising population and economic growth. Energy demand for climate control will triple by 2050, consuming more energy than that currently consumed altogether by the United States, the European Union and Japan. This increase in energy will directly impact water consumption, either to directly cool a condenser of an equipment or to serve indirectly as a basis for energy sources such as hydroelectric power that feed these heating, ventilation and air conditioning (HVAC) systems. Knowing the unique and growing importance of water, a new index, Total Water Impact (TWI) is presented, which allows a holistic comparison of the impact of water use on water, air and evaporative condensation climate systems. 200 and 500 TON (tons of refrigeration) air-cooled and water-cooled systems are theoretically compared to evaluate the general water consumption level. The TWI index is higher in the smallest water condensing system. That is, holistically, water consumption is higher in the water condensing system than in the air condensing system. Thus, this index provides a new insight about energy consumption and ultimately, about sustainability.

Keywords: HVAC; water-cooled condenser; air-cooled condenser; evaporative; TWI

---

1. Introduction

According to the Worldwide Fund for Nature (WWF) [1], there are more than 326 quintillion liters of water on earth. Less than 3% of all this water is freshwater, and of this amount, more than two thirds are in polar ice caps and icebergs. Also, according to the Brazilian Ministry of Environment [2], freshwater is not evenly distributed across the globe. Its distribution depends essentially on the ecosystems that make up the territory of each country. According to the International Hydrological Program of the United Nations Educational, Scientific and Cultural Organization (UNESCO), in South America, there is 26% of the total available freshwater on the planet and only 6% of the world population. Asia has 36% of total water and is home to 60% of the world's population (see Figure 1) [3].

Daily water consumption varies widely around the globe. In addition to site availability, average water consumption is strongly related to the country's development and economic levels. Each person needs, on a daily basis, at least 40 liters of water to drink, shower, brush their teeth, wash their hands and cook. However, UN data [4] indicate that a European, who has 8% of the world's freshwater on its territory, consumes an average of 150 liters of water per day. An Indian only consumes 25 liters a day. According to UNESCO estimates to 2025, based on the current rate of population growth and the failure to establish sustainable water consumption, this will lead to a human consumption up to 90%, leaving only 10% for other living beings on the planet [3].

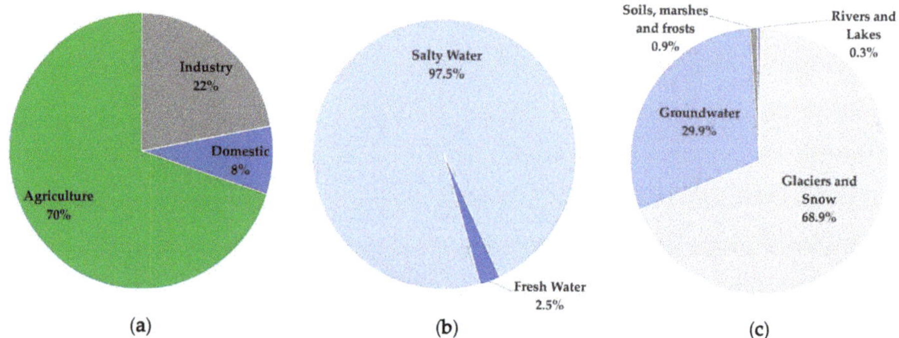

Figure 1. World water consumption and water distribution [3]. (**a**) Water consumption, (**b**) global total water, (**c**) 2.5% global total freshwater.

Brazil, with an area of approximately 8,514,876 km^2 [5] and more than 190 million inhabitants, is today the fifth country in the world both in territorial extension and population. Due to its continental dimensions, Brazil has great contrasts related not only to climate, original vegetation and topography, but also to population distribution and economic and social development, among other factors.

Overall, Brazil is a privileged country in terms of volume of water resources, as it houses 13.7% of the world's freshwater. However, the freshwater availability is not uniform. As shown in Figure 2, over 73% of the country's available freshwater is in the Amazon basin, which is inhabited by less than 5% of the population.

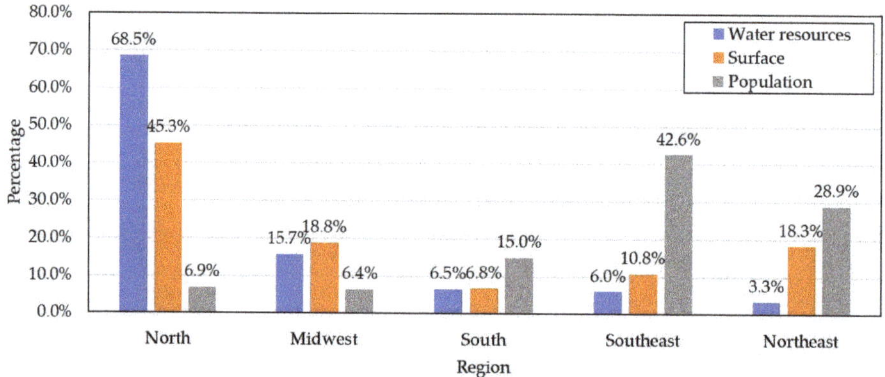

Figure 2. Distribution of water resources in the regions of Brazil [6].

Only 27% of Brazil's water resources are available to other regions, where 95% of the country's population lives. Not only is water availability non-uniform, but the supply of treated water reflects the contrasts in the development of Brazilian states. While in the Southeast region, 87.5% of households are served by a water distribution network, in the Northeast the percentage is only 58.7%.

Brazil also has high water waste: from 20% to 60% of treated drinking water is lost in distribution, depending on the conservation conditions of the supply networks. In addition to these water losses on the way between the treatment plants and the consumer, waste is also high at home, for example, involving the time required to take a bath, how the bath is taken, the large volume of the toilet water deposit, washing dishes with running water, the use of the hose as a broom to clean sidewalks, how cars are washed, etc. [6].

According to the International Energy Agency (IEA), there are over 500 million air conditioners in the world. Additionally, there are 2.8 billion people living in the hottest places in the world and only

8% have air conditioning. The numbers of air conditioners will increase from 1.6 billion in 2018 to 5.6 billion by 2050. Thus, 10 new air conditioners will be sold every second. The power supply for air conditioning by 2050 will be equivalent to the current USA, European Union and Japan electrical demand. Due to the importance of energy efficiency and water use in heating, ventilation and air conditioning (HVAC) systems, the aim of this paper is to generate a balance of these two resources [7].

HVAC systems are among the major consumers of freshwater. According to the Brasilian Association of Refrigeration, Air Conditioning, Ventilation and Heating (ABRAVA—Associação Brasileira de Refrigeração, Ar Condicionado, Ventilação e Aquecimento), the power capacity installed in Brazil will reach 60 million tons of refrigeration by 2029 [8]. Knowing the importance and connection of HVAC systems and water consumption, the aim of this work is to create an index to measure direct and indirect water consumption in HVAC systems.

2. Power Generation and Water Consumption in Brazil, Portugal and the USA

Water is present in over 70% of the biosphere. It is an essential resource for power plants. A survey was conducted in the United States of America (USA) by the National Renewable Energy Laboratory (NRL) on water consumption, extracting factors for electricity generating technologies. This data was not found for other countries. The USA energy matrix and the water consumption per kWh generated by different power sources is shown in Figure 3 [9].

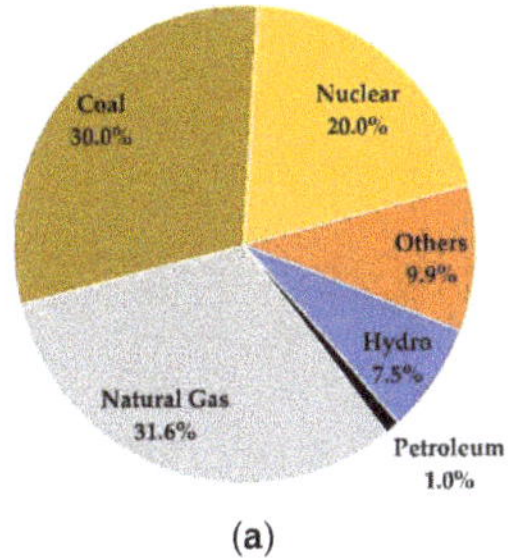

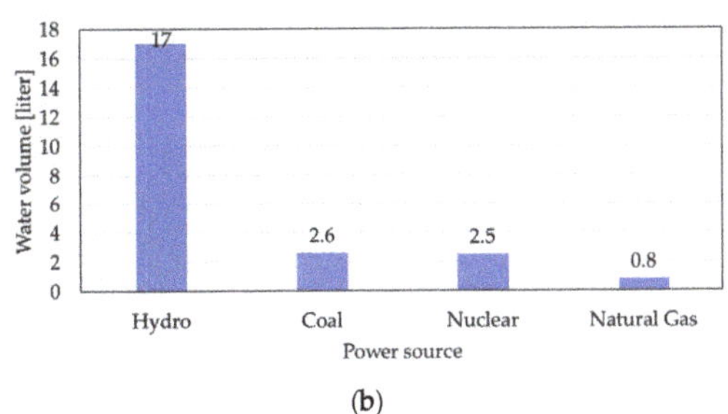

Figure 3. USA energy matrix and water consumed to generate 1 kWh of electricity. (**a**) USA energy matrix (ASHRAE—American Society of Heating, Refrigerating and Air-Conditioning Engineers), (**b**) water consumed to generate 1 kWh of electricity [9].

According to Portuguese Renewable Energy Association (APREN—Associação Portuguesa de Energias Renováveis), renewable energy systems generated 17.2 GWh during the first half of 2018, which represents 61% of the mainland Portugal's electricity production (28.2 GWh). This result is largely driven by the increased availability of resources, especially water and wind [10]. In that six month period, it is also worth highlighting a set of 623 non-consecutive hours (equivalent to 26 days) in which renewable electricity alone was enough to supply Portuguese electricity consumption (Figure 4).

In Brazil, according to the national energy balance of 2018 [11], the distribution of electric power generation has the distribution shown in Figure 5. Hydroelectric power generation is predominant, with a share of 65.2%. In Brazil, there is a tax incentive for the construction of small hydroelectric plants, thus the total share of small hydroelectric and hydroelectric plants is very large. The share of solar and wind plants in electricity production in Brazil is still small.

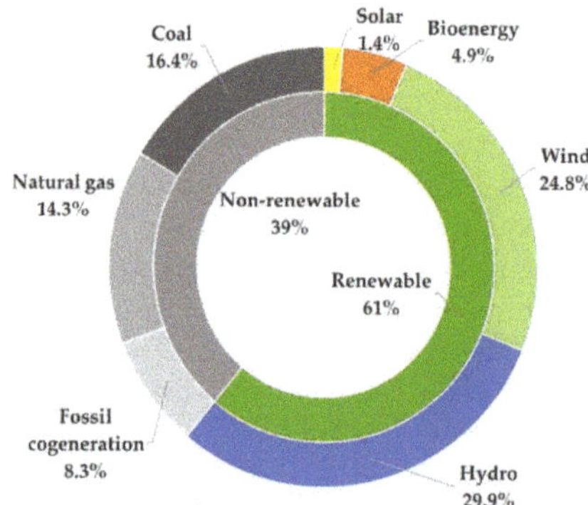

Figure 4. Portugal's electric energy matrix [10].

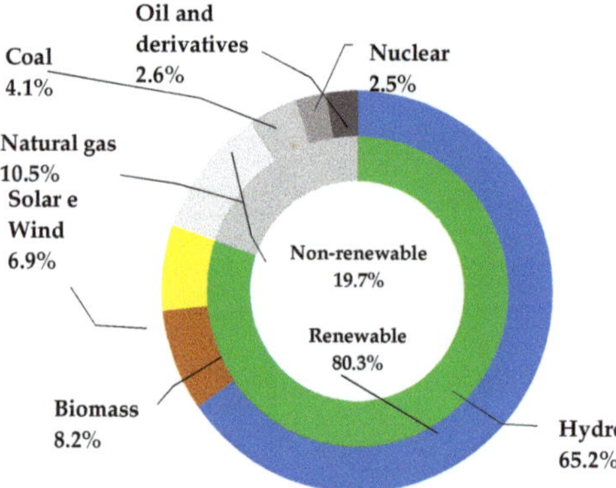

Figure 5. Brazil's electric energy matrix [11].

Figure 6 shows that the hydroelectric is the highest in Brazil, USA and Portugal, respectively. Although coal, biomass and oil have different amounts of pollution rates for power generation through thermoelectricity, the principle of heat transformation for energy and the need of water for steam generation are similar, thus they are considered together.

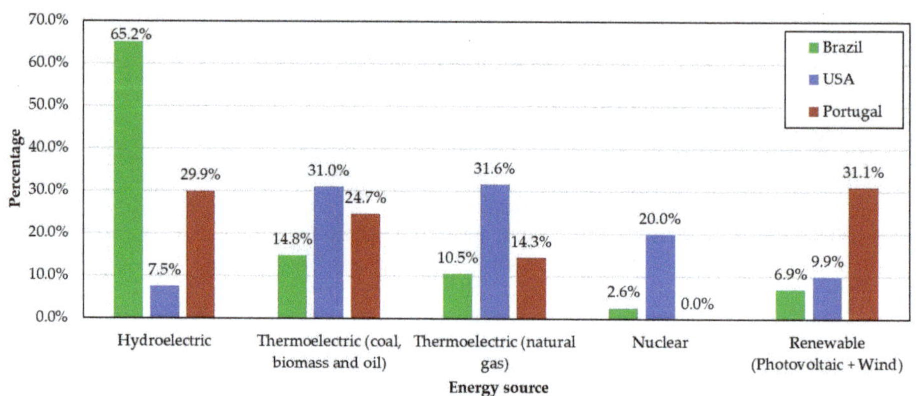

Figure 6. Electricity Generation types in Brazil, USA and Portugal.

Due to the lack of studies on the amount of water used per kWh generated in energy sources such as photovoltaic and wind, even knowing that water is needed to clean the photovoltaic panels and wind turbine propellers, a null water demand will be considered. Equation (1) provides the number of liters of water per kWh generated in the country required for each of these sources.

$$LAG(country) = (\%ghca) + (\%gtca) + (\%gtgnca) + (\%gnca) + (\%goca) \tag{1}$$

where LAG(country) is liters of water per kWh generated in the country, %ghca is % hydroelectric generation–water consumption per kWh hydroelectric, %gtca is % coal and oil thermoelectric generation–water consumption per thermoelectric kWh, %gtgnca is % thermoelectric generation natural gas–water consumption per thermoelectric kWh, %gnca is % nuclear generation–water consumption per nuclear kWh and %goca is % generation from other sources (renewable)–water consumption per kWh others.

$$LAG(Brazil) = (0.652 \times 16.2) + (0.148 \times 2.49) + (0.105 \times 0.72) + (0.206 \times 2.412) + (0.069 \times 0)$$
$$LAG(Brazil) = 11.071 \, L/kWh$$
$$LAG(USA) = (0.075 \times 16.200) + (0.3001 \times 2.490) + (0.316 \times 0.720) + (0.200 \times 2.412) + (0.99 \times 0)$$
$$LAG(USA) = 2.697 \, L/kWh$$
$$LAG(Portugal) = (0.299 \times 16.2) + (0.247 \times 2.49) + (0.143 \times 0.72) + (0 \times 2.412) + (0.311 \times 0)$$
$$LAG(Portugal) = 5.561 \, L/kWh$$

Due to the high hydroelectric rate in Brazil, water consumption per kWh generated is 410.5% higher than the USA average value and 50.22% compared to Portugal, as shown in Figure 7.

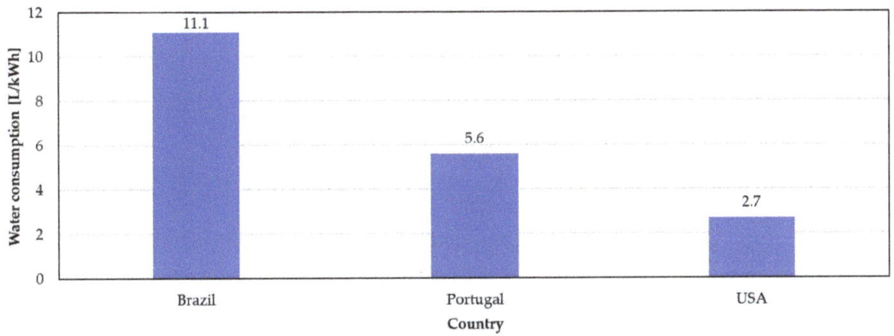

Figure 7. Water consumption (L/kWh).

With the fast development of the global economy, depletion of water resources is becoming an environmental issue of the utmost concern worldwide. The United Nations World Water Development Report published by UNESCO [12] indicates that water for current uses is becoming scarce and is leading to a water crisis. The effects that a sector can have on the environment are nowhere more visible than in the building industry. Building construction and its operations draw heavily on water from the environment. Growth in urban water use has caused a significant reduction of water tables and requires large projects that siphon supplies away from agriculture. Water used to operate buildings is a significant component of every nation's water consumption. However, this is not the only form of water consumed throughout a building's life cycle. Water is also consumed in the extraction, production, manufacturing and delivery of materials and products to site, and the actual on-site construction process, named as "embodied" water [13]. Ilha et al. [14] observed that water conservation technologies and strategies are often the most overlooked aspects of a whole-building design strategy. However, the planning for various water uses within a building is increasingly becoming a high priority, in part because of the increasing recognition of the water savings that can be reached through the implementation of water saving initiatives.

The LEED (Leadership in Energy and Environmental Design) already emphasizes the need for reducing water use, and states that buildings play a huge role in this effort. Buildings are the third biggest user of potable water in the USA. However, water efficiency considers only 12 of 110 points for LEED certifications (LEED v4.1). This certification has three prerequisites (outdoor water use reduction, indoor water use reduction, building level water metering) and credits for outdoor water use reduction, indoor water use reduction, cooling tower and process water use—water metering. Thus, almost 11% of LEED certification is about water efficiency. Because of the importance of water efficiency in large buildings, the use of water-cooled condenser systems may seem impractical, but the energy savings in these systems associated with less space and noise level are essential.

In simulations, as the wet bulb temperature is smaller than the dry bulb temperature, it is normal for water-cooled condensers to consume less energy than air-cooled condensers but losing water efficiency points in a LEED certification is not a desire of the certifier of green building.

Therefore, the purpose included in this paper is to create an index that can measure water directly (evaporation in water-cooled condenser) and indirectly (water needed for the production of the input energy of the system) [15].

3. Total Water Impact (TWI)

The Total Water Impact (TWI) index is given by the amount of the holistically required water for air conditioners over the life of the equipment (Figure 8). It was developed as a new metric for building certification.

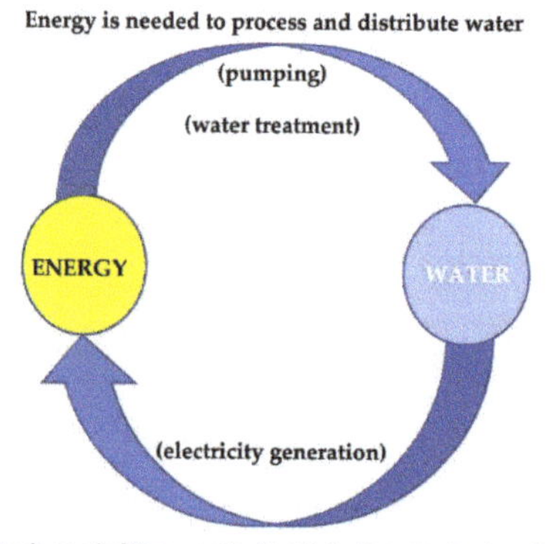

Figure 8. Energy–water nexus (adapted from Reference [11]).

The interdependencies between water and energy systems are clear and are becoming more prominent as overall development requires more resources, while their overuse and climate change impact make some resources scarcer. At the macro level, water is used at all stages of energy production and electricity generation (including renewable energy). Energy is required to extract, transport and supply water, and to treat wastewater prior to its return to the environment. At the micro level, the water–energy nexus is an important consideration for the HVAC community, dedicated to equipment design and selection and design of systems, as well as in construction operations.

One way to analyze the energy efficiency of large equipment is given by efficiency indicators, particularly, the non-standard part load value (NPLV). Specifically, NPLV is derived from the integrated part load value (IPLV). Both are used to evaluate chiller efficiency under different load conditions. The calculation formulas are the same. The only difference between IPLV and NPLV is that IPLV is calculated according to the condition specified in AHRI_Standard_550-590, while NPLV is calculated according to the condition of the location where the equipment is installed [16].

Thus, taking into account the above considerations, the simplified TWI is given by Equation (2):

$$\text{TWI} = (\text{ATL·NPLV·ELS·RRW}) + (\text{ATL·ELS·WUTR}) \tag{2}$$

where TWI is Total Water Impact, in m^3, ATL is Annual Thermal Load, in TON/year, NPLV is non-standard part load value, in kW/TON, ELS is equipment lifespan, in years, RRW is region-specific flow rate of water, in m^3/kW generated and WUTR is water used by TON, in m^3/TON of evaporation, drag and purge. Henceforth, the indicated TON unit refers to ton of refrigeration (1 TON = 3.5 kW).

The NPLV must be summed by TON from a Condensation Pump and Tower Fan. The result will be the total impact of holistic water use over the lifespan of the system.

There is already an indicator that is known as TEWI (Total Equivalent Warming Impact), which is the sum of direct and indirect emissions through refrigerant gas losses and life cycle energy consumption.

TWI is intended to measure the direct and indirect water consumption in the life cycle; therefore, the difference between TEWI and TWI is that TEWI measures GWP (Global Warming Potential), while TWI measures direct and indirect water consumption of the HVAC system in the life cycle. It is important in "Green Building" to place operational characteristics as part of the indices, as stated by Al-Ghamdi [17]. Like TEWI, TWI has advantages of using energy matrix characteristics in the calculations.

4. Water and Air-Cooled Comparison

In this section, the values of the air and water condensation systems' indices are determined and compared. Figure 9 shows the type of devices required when using air- or water-cooled chillers.

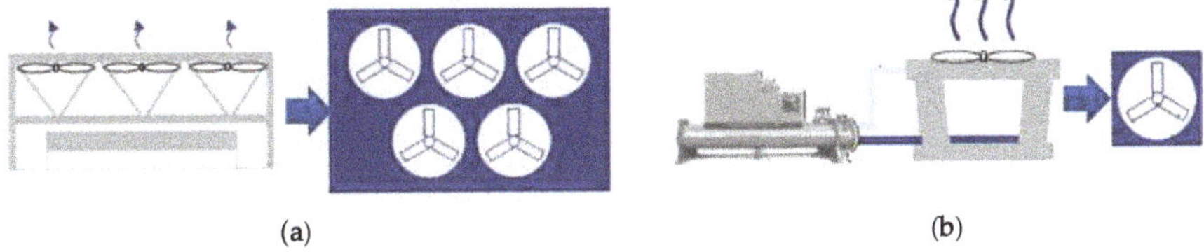

(a)   (b)

Figure 9. Comparison of (**a**) air- and (**b**) water-cooled systems [18].

The air-cooled system (Figure 9a) is a design based on dry bulb temperature:

- Larger occupied area (more surface area).
- Higher noise level.
- Higher energy consumption: lower efficiency.
- Does not consume water on site (without evaporative cooling).

The water-cooled system (Figure 9b) is a design based on wet bulb temperature:

- Smaller occupied area (usually requires equipment room).
- Low noise level.
- Lower power consumption: higher efficiency.
- Consumes water (evaporative cooling).

To compare and use in an example, air and water condensation systems for a total capacity of 200 TON with the characteristics and specifications shown in Table 1 were selected. A Brazilian air condensation chiller (Samurai screw compressor type, manufactured in Brazil) and condensation (imported) water chiller (RTHD model, manufactured in USA) were compared. Simulations were performed using the annual thermal load, the NPLV (given already with dynamic partial loads).

The average "ASHRAE" of water consumption per TON is also an input for the simulation of the water condensation system. The simulation period considered was 1 year. Life cycle of the equipment was set for 15 years. Calculations were based on ASHRAE NPLV formulas. Specifically, the thermal load was assumed, because each enclosure has different thermal loads. The TWI for each option (air- and water-cooled), according to Equation (2), is:

$\text{TWI}_{\text{air cooled}} = (657{,}000 \text{ TON/year} \times 1.06 \text{ kW/TON} \times 15 \text{ years} \times 0.011071 \text{ m}^3/\text{kW}) + (657{,}000 \text{ TON/year} \times 15 \text{ years} \times 0 \text{ m}^3/\text{TON})$

$\text{TWI}_{\text{air cooled}} = 115{,}650.987 \text{ m}^3$

$\text{TWI}_{\text{water cooled}} = (657{,}000 \text{ TON/year} \times 0.36 \text{ kW/TON} \times 15 \text{ years} \times 0.011071 \text{ m}^3/\text{kW}) + (657{,}000 \text{ TON/year} \times 15 \text{ years} \times 0.0072 \text{ m}^3/\text{TON})$

$\text{TWI}_{\text{water cooled}} = 110{,}233.694 \text{ m}^3$

Table 1. 200 TON system parameters.

Parameter	Air Condensation	Water Condensation
Full Load Chiller Efficiency (kW/TON)	1.4	0.65
NPLV partial load efficiency chiller (kW/TON)	1.06	0.36
System total capacity (TON)	200	200
Average annual thermal load (TON)	150	150
Condensation Pump power (kW input)	0	6.2664
Tower Fan power (kW input)	0	4.1776
Hours of operation in the year (hour/year)	4380	4380
Annual thermal load (TON/year)	657,000	657,000
Water demand per m^3/kW generated region	0.011071	0.011071
Water Consumption Index/TON, Tower (m^3/TON)	0	0.0072
TWI (m^3)	115,650.987	110,233.694

As another example, an air and water condensation system for a total capacity of 500 TON with NPLV [17] will be compared with the characteristics shown in Table 2. The equipment of 200 and 500 TON used the same simulation with the energy consumption data of manufacturers (200 TR) or ASHRAE 90.1 (500 TR). Again, a life cycle of the equipment equal to 15 years was set. The chillers of 200 and 500 TON have different capacities and different energy efficiency, but to exchange heat of 1 TON, both have the same consumption of evaporation water. The TWI for each option (air and water condensation), according to Equation (2), is:

$\text{TWI}_{\text{air cooled}} = (1{,}270{,}200 \text{ TON/year} \times 0.745 \text{ kW/TON} \times 15 \text{ years} \times 0.011071 \text{ m}^3/\text{kW}) + (1{,}270{,}200 \text{ TON/year} \times 15 \text{ years} \times 0 \text{ m}^3/\text{TON})$

$\text{TWI}_{\text{air cooled}} = 166{,}217.419 \text{ m}^3$

$\text{TWI}_{\text{water cooled}} = (1{,}270{,}200 \text{ TON/year} \times 0.38 \text{ kW/TON} \times 15 \text{ years} \times 0.011071 \text{ m}^3/\text{kW}) + (1{,}270{,}200 \text{ TON/year} \times 15 \text{ years} \times 0.0072 \text{ m}^3/\text{TON})$

$\text{TWI}_{\text{water cooled}} = 221{,}963.639 \text{ m}^3$

In both air and water condensing systems with capacities of 200 or 500 TON, the water consumption per kWh generated, and the water consumption index/TON, Tower (m^3/TON) have the same values.

Using the TWI calculation, which aims to understand water consumption holistically, the need for air conditioning equipment during its lifetime is determined.

Table 2. 500 TON system parameters.

Parameter	Air Condensation	Water Condensation
Full Load Chiller Efficiency (kW/TON)	1.237	0.585
NPLV partial load efficiency chiller (kW/TON)	* 0.745	* 0.38
System total capacity (TON)	500	500
Average annual thermal load (TON)	290	290
Condensation Pump power (kW input)	0	15.666
Tower Fan power (kW input)	0	10.444
Hours of operation in the year (hour/year)	4380	4380
Annual thermal load (TON/year)	1,270,200	1,270,200
Water demand per m^3/kW generated region	0.011071	0.011071
Water Consumption Index/TON, Tower (m^3/TON)	0	0.0072
TWI (m^3)	166,217.419	221,963.639

* Efficiency based on ASHRAE 90.1-2016. Specifically, in this case IPLV = NPLV because it is not a specific city [18].

Checking the TWI results in Tables 1 and 2, specifically in the hypothetical comparison of the 200 TON air (Brazilian) and water (USA imported) systems, the TWI index was lower in the water condensation type system. That is, holistically, water consumption was lower in the water-cooled system than in the air-cooled system.

In the system with 500 TON cooling power, according to ASHRAE 90.1 of 2016 [19], considering air- and water-cooled chillers, the highest value of the TWI index was obtained in the water-cooled chiller. Larger chillers were used because normally ASHRAE 90.1 uses package air conditioning, splits, for "base line" in small buildings and not chillers.

It is understandable that designers are concerned about the efficiency of the chiller. Several factors such as poor maintenance, malfunction, improper sizing, etc., affect this efficiency. Industry stakeholders know that any element that enhances any aspect of cooler efficiency can have a huge impact. In the case of the 500 TON power system, where the 2013 ASHRAE 90.1 values were used as the NPLV source, the TWI index value was lower in the air-cooled chiller.

As new standards emerge, chiller efficiency is becoming increasingly important: water-cooled magnetic bearing chillers in warm temperatures have low NPLV, certainly leading in the future to lower TWI indexes than air-cooled chillers [18].

5. Conclusions

The Total Water Impact (TWI) index provides a holistic view of water consumption over a period of time. With this methodology, it can be observed that sometimes an air-cooled system has higher water consumption than a water-cooled system.

Water consumption has not yet been considered in cleaning the air condenser coil (even because the cleaning method for water condensing systems is using a brushing system). Likewise, the cleaning water for wind and solar energy sources was not included as this value has not yet been found in technical or scientific publications.

Two examples were used to demonstrate the validity of the TWI index. It was concluded that the TWI shifts from water-cooled systems to air-cooled systems as refrigeration power increases. Thus, the holistic use of water decreases for air-cooled chillers as the refrigeration power increases, while the opposite condition occurs for water-cooled chillers. That is, the holistic use of water increases with the refrigeration power required. However, it must be highlighted that the NPLV values of systems with magnetic bearing compressors in water-cooled chillers are getting smaller. The same methodology can be used in systems with variable refrigerant volume (VRF) with water and air condensation. In a future work, geothermal systems and dry coolers will be simulated for TWI.

Among the indicators for choosing the best cooling system and even for LEED-certified systems, the TWI can be a quality indicator. It can also be considered a pro rata TWI per TON (result divided by the cooling load in the system lifetime) for decision making. An advantage of this new index is to use the same formula in the consumption of water and energy connected by the characteristics of the region in which the systems are located. In the future, it may be part of LEED or AQUA assessment. Therefore, TWI pro rata could be an index associated to both water and energy assessments.

Author Contributions: Conceptualization, A.F.S. and P.D.G.; methodology, A.F.S.; validation, A.F.S and H.J.L.S.; formal analysis, A.F.S. and P.D.G.; investigation, A.F.S.; resources, H.J.L.S.; data curation, A.F.S. and P.D.G.; writing—original draft preparation, A.F.S., H.J.L.D.S.; writing—review and editing, P.D.G.; visualization, H.J.L.D.S.; supervision, P.D.G.; project administration, A.F.S. All authors have read and agreed to the published version of the manuscript.

Funding: This research received no external funding.

Conflicts of Interest: The authors declare no conflict of interest.

Abbreviations

ABRAVA	Brazilian Association of Refrigeration, Air Conditioning, Ventilation and Heating
AHRI	Air conditioning, Heating and Refrigeration Institute. Actually, does the rating of air conditioners and refrigeration units
APREN	Portuguese Renewable Energy Association
AQUA	Project to Achieve High Environmental Quality in New Projects
ASHRAE	American Society of Heating, Refrigeration & Air Conditioning Engineers
ATL	Annual Thermal Load, in TON/year
ELS	Equipment lifespan, in years
ghca	Hydroelectric generation–water consumption per kWh hydroelectric, in L/kWh
gnca	Nuclear generation–water consumption per nuclear kWh, in L/kWh
goca	Generation from other sources (renewable)–water consumption per kWh, in L/kWh
gtca	Coal and oil thermoelectric generation–water consumption per thermoelectric kWh, in L/kWh
gtgnca	Thermoelectric generation natural gas–water consumption per thermoelectric kWh, in L/kWh
GWP	Global Warming Potential
HVAC	Heating, Ventilating and Air Conditioning
IEA	International Energy Agency
IPLV	Integrated Part Load Value
LAG	Liters of water per kWh generated in the country, in L/kWh
LEED	Leadership in Energy and Environmental Design
NPLV	Non-standard part load value, in kW/TON
RRW	Region-specific flow rate of water, in m^3/kW generated
TEWI	Total Equivalent Warming Impact
TON	Tons of Refrigeration (1 TON = 3.5 kW)
TWI	Total Water Impact, in m^3
UNESCO	United Nations Educational, Scientific and Cultural Organization
USA	United States of America
WUTR	Water used by TON, in m^3/TON of evaporation, drag and purge
WWF	Worldwide Fund for Nature

References

1. WWF. Programa Hidrológico Mundial. Available online: https://www.wwf.org.br/wwf_brasil/wwf_mundo/ (accessed on 30 March 2020). (In Portuguese)
2. MMA. Água. Ministério de Meio Ambiente (MMA). Available online: https://www.mma.gov.br/estruturas/sedr_proecotur/_publicacao/140_publicacao09062009025910.pdf. (accessed on 30 March 2020). (In Portuguese)

3. WINS. Water Information Network System (WINS) by the International Hydrological Programme ofUnited Nations Educational, Scientific and Cultural Organization (UNESCO). Available online: http://ihp-wins.unesco.org/ (accessed on 30 March 2020).

4. UN. World Water Development Report 2019. Available online: https://www.unwater.org/publications/world-water-development-report-2019/ (accessed on 30 March 2020).

5. IBGE. *Anuário Estatístico*; Instituto Brasileiros de Geografia e Estatística–IBGE: Rio de Janeiro, Brazil, 2017. (In Portuguese)

6. ANA. *Conjuntura dos Recursos Hídricos no Brasil*; Agência Nacional de Águas—ANA: Brasília, Brazil, 2013. (In Portuguese)

7. IEA. Data and Statistics. Available online: https://www.iea.org/statistics/?country=WORLD&year=2016&category=Energy%20supply&indicator=TPESbySource&mode=chart&dataTable=BALANCES (accessed on 30 March 2020).

8. ABRAVA. *Seminário Brasileiro de Etiquetagem em Eficiência Energética para sistemas AVAC*; ABRAVA: São Paulo, Brazil, 2018. (In Portuguese)

9. NREL. National Renewable Energy Laboratory. Available online: https://www.nrel.gov/ (accessed on 30 March 2020).

10. APREN. *Boletim Energias Renováveis. Portugal*; 1° semestre de; APREN: Lisboa, Portugal, 2018; Available online: https://www.apren.pt/contents/publicationsreportcarditems/12-boletim-energias-renovaveis-2018.pdf (accessed on 30 March 2020). (In Portuguese)

11. EPE. Balanço Energético Nacional. Available online: http://epe.gov.br/pt/publicacoes-dados-abertos/publicacoes/balanco-energetico-nacional-2018 (accessed on 13 September 2019). (In Portuguese)

12. UNESCO. *The United Nations World Water Development Report 2019: Leaving No One Behind*; UNESCO World Water Assessment Programme; United Nations Educational, Scientific and Cultural Organization (UNESCO): Paris, France, 2019.

13. Akadiri, P.O.; Chinyio, E.A.; Olomolaiye, P.O. Design of a Sustainable Building: A Conceptual Framework for Implementing Sustainability in the Building Sector. *Buildings* **2012**, *2*, 126–152. [CrossRef]

14. Ilha, M.S.O.; Oliveira, L.H.; Gonçalves, O.M. Environmental assessment of residential buildings with an emphasis on water conservation. *Build. Serv. Eng. Res. Technol.* **2009**, *30*, 15–26. [CrossRef]

15. LEED v4.1. *Building Design and Construction. LEED_v4.1_BDC_Rating_System_1.2020*; U.S. Green Building Council: Washington, DC, USA, 2020.

16. AHRI_Standard_550-590. *Performance Rating of Water-Chilling and Heat Pump Water-Heating Packages Using the Vapor Compression Cycle*; Air-Conditioning, Heating, and Refrigeration Institute (AHRI): Arlington, TX, USA, 2015.

17. Al-Ghamdi, S.G.; Bilec, M.M. *Life Cycle Thinking and the LEED Rating System: Global Perspective on Building Energy Use and Environmental Impacts*; Civil and Environmental Engineering, University of Pittsburgh: Pittsburgh, PA, USA, 2015.

18. ASHRAE. *ASHRAE Energy Water Nexus–Denver*; American Society of Heating, Refrigeration & Air Conditioning Engineers (ASHRAE): Baltimore, MD, USA, 2016.

19. *ASHRAE/ANSI/IES Standard 90.1-2016-energy Standard for Buildings except Low-Rise Residential Buildings*; American Society of Heating, Refrigeration & Air Conditioning Engineers (ASHRAE): Atlanta, GA, USA, 2016.

MDPI

St. Alban-Anlage 66

4052 Basel

Switzerland

Tel. +41 61 683 77 34

Fax +41 61 302 89 18

www.mdpi.com

Energies Editorial Office

E-mail: energies@mdpi.com

www.mdpi.com/journal/energies